The Alchemy of Empire

The Alchemy of Empire

Abject Materials and the Technologies of Colonialism

Rajani Sudan

FORDHAM UNIVERSITY PRESS

NEW YORK 2016

Fordham University Press has no responsibility for the persistence or accuracy of URLs for external or third-party Internet
websites referred to in this publication and does not guarantee
that any content on such websites is, or will remain, accurate or
appropriate.

Fordham University Press also publishes its books in a variety of
electronic formats. Some content that appears in print may not
be available in electronic books.

Visit us online at www.fordhampress.com.

Library of Congress Cataloging-in-Publication Data

Names: Sudan, Rajani.
Title: The alchemy of empire : abject materials and the technologies
    of colonialism / Rajani Sudan.
Description: New York : Fordham University Press, 2016. |
    Includes bibliographical references and index.
Identifiers: LCCN 2015042059 (print) | LCCN 2015046367
    (ebook) | ISBN 9780823270675 (hardback) | ISBN
    9780823270682 (paper) | ISBN 9780823270699 (ePub)
Subjects: LCSH: Technology transfer—History—18th century. |
    Europe—Civilization—Oriental influences—History—
    18th century. | Enlightenment. | Technology in literature. |
    Imperialism—History—18th century. | Great Britain—
    Colonies—History—18th century. | BISAC: LITERARY
    CRITICISM / European / English, Irish, Scottish, Welsh. |
    HISTORY / Social History.
Classification: LCC T174.3 .S79 2016 (print) | LCC T174.3
    (ebook) | DDC 338.9/26—dc23
LC record available at http://lccn.loc.gov/2015042059

Printed in the United States of America

18 17 16   5 4 3 2 1

First edition

for Mum

CONTENTS

Mud, Mortar, and Empire

> Under the connecting feeling of tropical heat and vertical sun-lights,
> I brought together all creatures, birds, beasts, reptiles, all trees and
> plants, usages and appearances, that are found in all tropical regions,
> and assembled them in China or Indostan. From kindred feelings,
> I soon brought Egypt and all her gods under the same law. I was stared
> at, hooted at, grinned at, chattered at, by monkeys, by paroquets, by
> cockatoos. I ran into pagodas: and was fixed, for centuries, at the
> summit, or in secret rooms; I was the idol; I was the priest; I was
> worshipped; I was sacrificed. I fled from the wrath of Brama through
> all the forests of Asia: Vishnu hated me: Seeva laid in wait for me.
> I came suddenly upon Isis and Osiris: I had done a deed, they said, which
> the ibis and crocodile trembled at. I was buried, for a thousand years,
> in stone coffins, with mummies and sphinxes, in narrow chambers at
> the heart of eternal pyramids. I was kissed, with cancerous kisses, by
> crocodiles; and laid, confounded with all unutterable slimy things,
> amongst reeds and Nilotic mud. I thus give the reader some slight
> abstraction of my oriental dreams . . .
>
> —THOMAS DE QUINCEY, *Confessions of an English Opium Eater*

The Alchemy of Empire focuses on eighteenth-century British representations
of India and the crucial ways in which India's technology, scientific practice, and epistemology informed European Enlightenment values and sociopolitical norms. The value of this Indian techne[1] was eventually forgotten
or lost both in the crises of value and representation that characterized the
early decades of the Restoration and in the eventual triumph of Whiggish
history that trumpeted the narrative of European exceptionalism.[2] These

imported practices, however, emerged in the world of literature. Its public place in the history of British science was effectively devalued and effaced by nineteenth-century imperialist ideology. However, it continued to inform literary representations of the long eighteenth century, and it remains stubbornly lodged in the archives of the Oriental and India Office Collections of the British Library today. Indian arts and sciences provided English poets and novelists a language to account for alterity. Authors as varied as Alexander Pope, Lady Mary Wortley Montagu, Daniel Defoe, Samuel Johnson, John Cleland, Jemima Kindersley, and Jane Austen engaged models of otherness produced by the cultural appropriation of Indian techne by early modern England. But how did such models become so readily available to such disparate writers with such different writerly ambitions?

Throughout the eighteenth century, members of the British East India Company stationed in India reported their discoveries of native scientific and technological practices to the Royal Society. Their representations of these practices often focused on the miraculous or marvelous nature of technological discoveries. Isaac Pyke, governor of St. Helena, writes on the manufacture of mortar in Madras that forms a "stucco-work" surpassing any known European composition, particularly "Plaister of Paris . . . in smoothness and beauty" as durable as "marble."[3] Robert Coult, a doctor stationed in Calcutta, describes the wondrous method of smallpox inoculation practiced by Bengali Brahmins at least a century before Lady Mary Wortley Montagu pleaded with British doctors to adopt this form of medical science, and almost two centuries before Edward Jenner established smallpox vaccination as a wholly English medical practice. Likewise the manufacture of ice is successfully produced by Indian technology. Having "never heard of any persons having discovered natural ice in the pools or cisterns or in any waters collected in the roads," Sir Robert Barker begs "permission to present to you with the method by which it was performed at Allahabad, Mootegil, and Calcutta." This methodology, stimulated by the "Asiatic whose principal study is the luxuries of life," would make profitable a short duration of cold to alleviate the intense heat of the summer season, making habitation—with an eye to permanent colonial settlement—more comfortable for European visitors.[4]

Helenus Scott, another doctor stationed in Bombay, writes to Joseph Banks of the Royal Society that he will include in the next bill of lading a sample of *caute*, a surgical cement that putatively reattaches severed limbs. This matter was no doubt apocryphal, as no concrete evidence exists that such a substance was sent to England or was even available in India, but that very fact attests to the ease with which East India Company "servants" were willing to believe in the marvelous capacities of a techne they didn't understand and to document such beliefs. Letters from the missionary Jesuits to Paris equally extol Indian arts and sciences as miraculous and find their way into the Royal Society's *Philosophical Transactions*. Writing to Father Le Gobien in Paris in December of 1709, Father Papin claims that India "furnishes Materials for Mechanic Arts and Sciences more than any other Country that I know of"; providing an example of this superior techne, Papin notes that the "Artisans here have wonderful Skill and Dexterity" and "excel particularly in making Linnen-Cloath which is of such fineness that very long and broad Pieces of it may easily be drawn thro' a small Ring." Papin declares "if you tare a piece of Muslin into two Pieces, and give it to one of their Fine-Drawers . . . it will be impossible for you to discover where it is joined, tho' you mark it on purpose to know it." Similarly, "they will place together so artificially the Pieces of Glass or China Ware, that one cannot perceive it was ever broken."[5] The superior quality of Indian textiles was hardly contested in the early eighteenth century. Railing against the current fashion for Indian fabrics, Daniel Defoe famously exhorted Britons to be wary about destroying the trade in their own woolen goods and warned against the insidiousness of fashion and taste.

> Women servants are now so scarce, that from thirty and forty shillings a year, their wages are increased of late to six, seven, nay, eight pounds per annum, and upwards; insomuch that an ordinary tradesman cannot well keep one; but his wife, who might be useful in his shop or business, must do the drudgery of household affairs; and all this because our servant-wenches are so puffed up with pride nowadays, that they never think they go fine enough: it is a hard matter to know the mistress from the maid by their dress; nay, very often the maid shall be much the finer of the two. Our woollen manufacture suffers much by this, for nothing but silks and satins will go down with our kitchen-wenches;

to support which intolerable pride, they have insensibly raised their wages to such a height as was never known in any age or nation but this.[6]

The difference between Defoe's anxieties and Father Papin's observations, however, has to do with techne. That is, while Defoe acknowledges the existence of "silks and satins" that the very "kitchen-wenches," with their "intolerable" wages (although Defoe attributes it to their "pride") now desire, the production of these textiles is something that Father Papin prizes.[7] That such production occurs almost incidentally, with the minimum of investment in costly machinery, augments Papin's admiration for the "wondrous" dexterity of Indian artisans:

> The Looms that their Weavers use do not cost them more: With these they sit in their Courts and Yards, or on the side of the High-way, and work those fine Stuffs that are so highly esteem'd over all the World.[8]

The global esteem Father Papin grants the "Stuffs" Indian weavers produce, even "on the side of the High-way," suggests a universal acknowledgment of this form of Indian techne, something Defoe grudgingly implies in his disquisition against silks and satins.

The long eighteenth century was a period characterized by the "entanglement of science with the making of materials and material knowledge," and this epistemological endeavor was largely the province of chemists.[9] If we delve further back into the early modern period, however, we uncover a similar epistemological drive structuring the efforts of the major alchemists of the fifteenth and sixteenth centuries; Paracelsus, Oswald Croll, Otto Tachenius, Johann Rudolph Glauber, Johann Kunckel, and Johann Joachim Becher were all impelled to identify the chemical origins of their preparations, which they used as remedies. Craftsmen and artisans as well as philosophers, these alchemists were also salesmen, and they traveled (especially in the case of Becher) rather extensively throughout Europe and England peddling their wares. The connection between the making of material and material knowledge, then, was also crucially contingent on trade. Often considered the "creator" of empiricism, Francis Bacon outlines his ideas of experimental history in *Novum Organon*, in which he draws on the "commerce of the mind with things" and calls on the necessity of attending to the mechanical arts. It turns out this text was less of an originary moment than an articulation of a movement that had been emerging from the beginning of the fifteenth century.[10]

This brief historical sketch draws on the much more detailed and nuanced study done by Ursula Klein and Wolfgang Lefèvre, and in many ways my argument about the relationship between materials and knowledge parallels theirs—with two critical differences. The first has to do with methodology and provenance. Klein and Lefèvre's methodology is historical and their focus is on the "history of the most significant scientific objects of classical chemistry—chemical substances—covering European chemistry" in the early modern period.[11] My method adds literary as well as archival documentation, but my thesis suggests that some of the "classical" Enlightenment scientific knowledge is not European in origin but emerges from a far wider circulation. British science in particular was shaped by its relation with Indian techne. I also argue that the paradigm alchemy furnished was not definitively discarded with the emergence of eighteenth-century experimental chemistry. In fact, the entire project of transforming materials into knowledge relies on an alchemical transmutation, one that Klein and Lefèvre address as follows:

> Materials were indeed transformed when they became objects of inquiry for academic chemists. Chemists invested them with new meaning, and sometimes even transformed their boundaries by splitting them into different kinds of substances. . . . Studying chemical substances as applicable materials, perceptible natural objects, and things that carry imperceptible features, eighteenth-century chemists' inquiries moved from the perceptible to the imperceptible dimension of substances, and vice versa.[12]

The transformation of material to knowledge follows an alchemical sublimation of base material into something with infinitely more value: gold, silver, health, knowledge, or enlightenment produced from the *lapis philosophorum*. Father Papin's enthusiasm for India is shaped, in part, by the wealth of techne available to be learned and transmitted back to France, but his representation of the weaver's skill is as concerned with the "mechanic art" as it is with the marvel of invisibility. Despite Papin's efforts to keep the tear in view or locate the flaw in the "China ware," he fails, and in so doing, mystifies that failure. This mystification itself later transmutes to abjection, which is underscored by anxiety. However, writing in the beginning of the eighteenth century, Father Papin articulates a regard for the mechanical arts of Indian craftsmen that borders on the sublime.[13]

British epistemology clearly was informed and constituted by the exchange of techne that accompanied its trade interests.[14] The East India Company correspondents I investigate were fascinated by marvels and by encounters with things that they understood as fabulous.[15] Many of the substances they address are neither exotic nor alien, yet these writers endow them—mortar, plaster, ice, surgical glue, cloth—with mystical properties: plaster that acquires the hardness and durability of marble, vast quantities of ice made from no enduring source of chill, glue that reattaches body parts without scarring, and cloth woven so finely that tears are impossible to detect. *The Alchemy of Empire* excavates the histories and genealogies embedded in these materials of empire building: how they are identified, mystified, and transformed from attribute to substance. I argue that the model of transformation British East India Company members deploy is drawn from the language of alchemy (itself a borrowed concept), a paradigm of reason that turns base material into something with infinitely more value. For example, the ingredients of mortar (lime, toddy, jaggery) magically transform to "marble." In turn, their letters demonstrate, forms of alien techne are sublimated through alchemical language into transparent, enlightened, scientific truths, as Helenus Scott's desire to have his reports published in the Royal Society's *Philosophical Transactions* in the "interests of science" makes fairly explicit.

This book's argument both engages and resists traditional concepts of Western modernity that assume scientific and technological progress developed exclusively in Enlightenment Europe. I specifically cross-examine the relationship between material objects and their social and semiotic power in relation to received ideas about the intellectual and political power of the European Enlightenment. Such received ideas include a sense that imperial European ascendancy of the early modern period was based on "Enlightenment" discoveries of new human interactions with the world. Bringing together material things and historical events helps uncover the buried history and meaning of imperialism. *The Alchemy of Empire* reads the event of imperialism in relation to the transformation of modes of analogical argument to analytical argument, something that characterizes the European Enlightenment and its historically special modality of imperialism. The recovery of analogies, however—material substances, modes of behavior, and cultural understandings—reveals a richness of the confrontation of Britain and Indian material and cultural life that received notions of the European Enlightenment obscure.

The production of ice, the manufacture of mortar, the practice of smallpox inoculation, and the fabrication of paper are imbricated in the general rubric of empire. Hence Pope's *Windsor-Forest*, a celebration of the Pax Britannica, and most commonly read (in postcolonial studies) as a catalogue of imperial spoils, is saturated with the same alchemical language that members of the British East India Company were using to describe their encounters with Indian science. Likewise Barker's accounts of ice making, written during the Little Ice Age and highly informed by Daniel Fahrenheit's scale, allows for a climatological reading of Johnson's *Preface to the Dictionary*, a text informed by metaphors of chill and toil that favorably position this British lexicography against the "soft bowers" of similar French and Italian efforts.

I want to emphasize that I am treating these practices and processes of production—mud, mortar, ice, smallpox matter, cloth, and paper—as *technologies* of colonialism and empire. While the study of material culture broadens the scope of empire building, and reading these substances as both materials and tropes helps us excavate histories that have been erased or forgotten, it is important to understand the work these substances do in the production of imperial ideology. Mortar isn't simply the sum of its parts but becomes a technology of building a city that later represents British imperial presence in India. Ice making becomes part of the technology of transforming an alien climate into one more recognizably English. Inoculation transfers knowledge through vectors of disease; paper is embedded with the marks of intellectual labor. In many ways my claim that substances function as technologies reflects the use Bruno Latour makes of the term "substance." The conventional historical move is to identify a substance that exists in the outside world, without history, that through human intervention becomes a "phenomena observed by the mind." Latour reminds us that substances have histories, most of them located in circulating referents that reify them as substances. Using Pasteur's ferment as an example, Latour argues:

> The ferment began as a series of attributes and *ended up being a substance*, a thing with clear limits, with a name, with obduracy, which was more than the sum of its parts. . . . A substance is more like the thread that holds the pearls of a necklace together than the rock bed that remains the same no matter what is built on it. In the same way that accurate reference qualifies a type of smooth and easy circulation, substance is a name that designates the *stability* of an assemblage.[16]

I want to think about the relationship between attribute and substance as alchemy. In other words, it is human intervention that reifies (the alchemical process) substance into observable phenomena in the case of Pasteur's ferment and it is human intervention that reifies the "series of attributes" that British East India Company members see in India into substance. In so doing, I don't mean to privilege attribute over substance or the reverse; rather, I want to uncover the complicated processes that structure the ways in which we see and represent the phenomenological world. Read in this way, then, the ontology of mortar moves from being more than a sum of its parts, more than the recipe itself, to being both the chain of signification and the domain of practice that eventually guarantees the stability of British hegemony. Teasing out the ineluctable relation between material objects and historical events becomes the means to describe the buried history and meaning of imperialism. The social and semiotic power of such materials uncovers a narrative radically different from received ideas about the ascendancy of the European Enlightenment, and the reigning paradigm upon which such power depends is alchemy.

More fundamental to empire building than mortar, however, is the humble place mud occupies: as barren desert or Edenic garden, mud furnishes the grounds for making imperial claims. Because it operates variously as a signifier of techne in *Robinson Crusoe*, a demonic harbinger of oriental doom for Thomas De Quincey, and then a few decades later, as a stalwart metonymy of England for administrators of the British East India Company in India, mud reflects a complicated history of Britain's imperium. Of course, there are semiotic differences between the "soil" that mysteriously supplies Crusoe with an abundance of European grain despite the tropical climate, the ooze with which De Quincey modifies as "Nilotic," and the glorious mud that Lord Lytton lauds in his letter to his wife as a specifically pastoral "English." These differences speak to the virtues of agrarian models of nationhood that yield moral and material produce against the monsters of depraved cultures. Yet close readings of these three examples clarify the overdetermined position this unassuming and overlooked material of empire-building occupies.

For example, toward the end of his confessional narrative, Thomas De Quincey describes a dream in which he finds himself the object of malevolent scrutiny by various figures he associates with China and "Indostan." Fleeing their wrath "through all the forests in Asia," De Quincey bumps up against icons of Egyptian mythology who also berate him for an unmentionable crime, and his dream concludes with a mud bath in which he is "kissed with cancerous kisses, by crocodiles . . . confounded with all unutterable slimy things, amongst reeds and Nilotic mud." Writing in the colorful language of opiated fantasy, De Quincey expresses what was by 1822—the year his *Confessions* were first published—well-established English xenophobic anxieties about the East (as the Nile was a staple of European fantasy even beyond Shakespeare's *Anthony and Cleopatra*). Renewed critical interest in De Quincey's corpus has examined the relation between exotic commodities—particularly the consumption of opium—and national, cultural, and authorial identity. De Quincey's ambivalent consumption of opium articulates a generalized national and cultural ambivalence toward eastern exoticism that the English viewed as at once horrifying and desirable. De Quincey and other nineteenth-century writers were complicit in fleshing out these oriental fantasies with images of opulence and decay.[17] De Quincey's anxieties about dissolution, his fears of losing the cultural outlines that define his identity as a specifically English opium eater, articulate his desires to erase those definitions with equal clarity as many of these studies demonstrate. If we disengage the dream's *terroir* from the social relations that De Quincey scholars have investigated, however, another aspect of empire emerges, one that has hitherto been overlooked in postcolonial studies of imperial politics: mud *tells us something*.

Embedded in the Nilotic mud that concludes De Quincey's fantasy is an ancient system of values long since forgotten by an entire nation of English subjects. De Quincey wallows in mud from Egypt, a civilization long associated with the origins of alchemy. Ancient Egypt distinguished itself from the surrounding *deshret*, the red lands of the desert, by naming itself *Kemet*: the "black land" of Nilotic mud. During the Coptic phase of linguistic development, "Egypt" was signified by the hieroglyph *Khmi*, transliterated into Greek as *Khemia*. *Al-kimia*, the Arab word for a system that transmutes gold and silver, has multiple cultural origins as the *Oxford English Dictionary* explains:

> it was afterwards etymologically confused with the like-sounding Gr., pouring, infusion, f.—pf. Stem of-to pour, cf. juice, sap, which seemed to explain its meaning; hence the Renascence spelling of *alchymia* and *chymistry*. Mahn (*Etym. Unt. 69*) however concludes, after an elaborate investigation, that Gr. was probably the original, being first applied to pharmaceutical chemistry, which was chiefly concerned with juices or infusions of plants; that the pursuits of the Alexandrian alchemists were a subsequent development of chemical study, and that the notoriety of these may have caused the name of the art to be popularly associated with the name of ancient Egypt, and spelt, as in Diocletian's decree. From the Alexandrians the art and name were adopted by the Arabs, whence they returned to Europe by way of Spain.

The concatenation of alchemy's provenance overdetermines De Quincey's choice of a final resting place.[18] Poised between ancient and modern, between Egypt and Greece, between East and West, De Quincey's desires to dissolve into one of the unutterably slimy creatures inhabiting the mud places him at a cultural fulcrum, one that reiterates the twin drives of desire and loathing. He has also—perhaps unwittingly—placed himself between two mutually constitutive epistemological systems: alchemy and chemistry. The confounding mud furnishes the basis for both systems of knowledge; from the primeval Nilotic slime rise the double "cradles" of civilization, Alexandrian and Greco-Roman. These civilizations in turn furnish political, aesthetic, and philosophical models that shape the development of Enlightenment Europe and provide the template for Linnean taxonomy defining Enlightenment science. Of course, this fantastic scenario is one produced by De Quincey's own application of pharmaceutical chemistry and hardly the stuff of scientific fact. Yet the image of Nilotic mud and the epistemological system it signifies together provide a provocative paradigm for reading the politics of empire, one that silences an insistently English voice and invokes instead a host of other forgotten voices.

Some fifty years after the publication of De Quincey's *Confessions*, Robert Bulwer-Lytton, the first earl of Lytton and viceroy of India, wrote a letter to his wife from Ootacamund in the Nilgiri mountains praising the local mud: "I affirm it to be a paradise. The afternoon was rainy and the road muddy but such beautiful *English* rain, such delicious *English* mud!" Lord Lytton's response to conditions that would probably be annoying in most circumstances articulates a heartfelt relief at discovering an English landscape in an alien land. Indian rain and Indian mud transform the heathen hill station into an Edenic English sanctuary, one that Thomas Macaulay has earlier described as an "English watering place" with "the vegetation of Windsor forest or Blanheim spread over the mountains of Cumberland." Unlike the mud that silences De Quincey, this mud inspires rhetoric charged with enthusiastic comparisons to England. Lord Lytton likens Ootacamund to a fantastic amalgamation of British (and Roman) sites: "Hertfordshire lanes, Devonshire towns, Westmoreland lakes, Scottish trout streams, and Lusitanian views."[19] Of course, the mud isn't English at all, but by the late nineteenth century, Asiatic mud furnished a very different fantasy than the ones provoking De Quincey's dreams. Macaulay and Lytton see an English countryside spread before them and not simply as a generalized landscape. In addition to recognizing the various parts of pastoral England in the Nilgiri mountains, they also identify monarchical seats—Windsor Forest, Blanheim Palace, and, arguably, the Roman Republic—thus suggesting their complete transference of England to India. De Quincey's nightmare takes place in the bucolic heart of the Lake District: Dove Cottage. Transported and deposited in China by the Malay, De Quincey flees, only to lose himself in the Nilotic mud where he becomes an unutterable creature like the rest of the slimy "things." Lord Lytton looks outside upon Indian terrain and sees England in the mud created by a rainy day, which he notes in his letter to his wife back in England, almost as if "to assimilate India to the English self, rather than figuring it as radically other," as Kavita Philip argues.[20]

What does mud tell us about empire building? An abject material, mud is nevertheless central to these fantasies of dominion. It has the capacity to stifle De Quincey's rather garrulous literary voice and to arouse dozens of fond comparisons to the English pastoral from the most prosaic administrators of colonial India. Mud furthermore has defined at least one epistemological system that privileges the sublimation of base material into something of greater value: it clearly performs a geographical alchemy for Lytton and Macaulay in transubstantiating an alien landscape to one that is profoundly familiar to them. A closer look at the relation between mud and sublimation may help identify the overdetermined place it occupies in literary representations of British imperial enterprise.

There is, perhaps, no more gratifying use of mud in the eighteenth-century novel than in Daniel Defoe's first volume of *Robinson Crusoe*. The ground in which Crusoe has planted barley and rice at first yields nothing, and perhaps, given the tropical heat hostile to the cultivation of European grain, it shouldn't. Clearly, Crusoe's crop is ideologically produced according to Protestant virtues of transplantation and agrarian abundance. Defoe is careful to provide a telling detail that complicates a mere ideological reading: after seeking "a moister piece of ground to make another trial in"—finding a more accommodating mud, as it were—Crusoe is rewarded with a small crop that establishes him as "master of [his] business."[21] Crusoe's agricultural success makes greater demands on his use of the local mud; he needs earthenware vessels in which to store his grain and cook his increasingly diversified meals. He finds clay out of which he makes an earthen "paste" with which to shape into vessels: "No joy at a thing of so mean a nature was ever equal to mine, when I found I had made an earthen pot that would bear fire."[22] In another instance of self-fashioning, Crusoe's plot and Crusoe's pot convert the alien territory onto which he has providentially landed into a replica of the English pastoral. Virginia Woolf comments on the prosody of this pot noting that

> Defoe, by reiterating that nothing but a plain earthenware pot stands in the foreground, persuades us to see remote islands and the solitudes of the human soul. By believing fixedly in the solidity of the pot and its earthiness, he has subdued every other element to his design; he has roped the whole universe into harmony. And is there any reason, we ask as we shut the book, why the perspective that a plain earthenware pot exacts should not satisfy us as completely, once we grasp it, as man himself in all his sublimity standing

against a background of broken mountains and tumbling oceans with stars flaming in the sky?[23]

Woolf homes in on the "earthiness" of this pot, accentuating its sturdiness: the earthen material is so basic and so *solid* that it stands in for "every other element." Moving from the "solidity" of Crusoe's pot to the "sublimity" of "man himself," Woolf engages an alchemical paradigm to recognize the ways in which horizons of alterity are brought within the grasp of an imperial "perspective."[24] The pot's very plainness, its mundane existence as one of the unsung workhorses of domestic labor, is the medium by which it sublimates into the invaluable material that brings "remote islands" and the "whole universe" into an imperial "harmony." But Woolf's totalizing standpoint is erroneous even if it reflects the historical moment of British self-perception. Although Defoe's novel has been read in contemporary studies of empire as a tour-de-force handbook of British colonialism, it's worth considering the symptomatic differences Defoe insists on.

When he first washes up on shore, Crusoe's wild reflections on island matters eventually culminate in his decision to spend the night in a great bushy tree where he can contemplate the melancholy fates that lie in store for him: to die of starvation or to be eaten by animals.[25] The next morning, however, all has changed. Descending from his "apartment" in the tree, he "look[s] about [him] again" and somehow makes peace with his new circumstance and adapts to the conditions imposed by the island.[26] Of course the difference in "prospects" from one day to the next is that the ship has drifted into closer proximity and may afford "something for my present subsistence." While fabricating a hybrid self-sustaining culture from stores in the foundered ship and island resources, Crusoe only replicates a semblance of an English agricultural and commercial landscape.

Nowhere is the fragility of this fantasy more evident than when he encounters another plain "earthen" print planted by deus ex machina into the foreground of his perspective: "It happened one day about noon going towards my boat, I was exceedingly surprised with the print of a man's naked foot on the shore, which was very plain to be seen in the sand."[27] Crusoe's accounts for the presence of this footprint range from one frenzied supposition

to another until he finally accepts the inevitable. What is interesting, however, is his reaction immediately after determining that the print is indeed
human. After years of proudly claiming his "country house" and "seacoast
house," in a few short minutes after seeing the footprint those boasts
evaporate, leaving him with little purchase on which to sustain his former
claims. He now returns to his "fortification, not feeling, as we say, the ground
I went on." Embodied in that footprint is a clear contestation to his island
tenancy; the jubilant self-image he has appropriated as his own suddenly vanishes as the ideological underpinnings that sustain his harmonious domestic fantasy dissolve into the sand. Hence his confusion as he hastens back to
the last remainder of his island property; home no longer, now only the last
line of defense. Incapable of even "feeling" the ground beneath him much
less being in possession of it, Crusoe's facility to read the landscape, initially
granted him by his arboreal apartment, abruptly disappears and the island
becomes newly strange: terrified, he looks behind him every two or three
steps, "mistaking every bush and tree, and fancying every stump at a distance to be a man."[28]

More than one hundred years separate Defoe's 1719 novel and De
Quincey's 1822 publication, and in that separation one reads the various
iterations of mud's representational capacity. In the case of both texts, mud
has the capacity to signify promise and horror. The same ground that yields
to Crusoe a wealth of agricultural and domestic bounty also inscribes the
real presence of native others—the "cannibals" of Crusoe's fears—whose
sovereignty is clearly established by their movement from mainland to island
while Crusoe, despite careful husbandry and efforts at chartering the
island, is still its prisoner. The Nilotic mud in which an ancient symbolic is
buried also harbors crocodiles and "unutterable slimy things." Grounded, as
it were, in these conflicting responses to mud is an alternative history to
Enlightenment accounts of European exceptionalism. The same mud that
offers a scientific paradigm upon which the principles of modern science
rested also provides concrete evidence of the presence of an autonomous
other. Thus mud performs alchemy of both types: it can transmute into an
abstraction of infinite value, and it can be demystified into its base state,
which demonstrates its utter necessity in the production of imperial power,
as we see in these three literary cases.

It may suffice to argue that until quite recently, the disavowal of Indian techne was largely a result of an ideological necessity to maintain imperial dominion, and that, with the collapse of British control over its colonies in the east and the recovery of forgotten histories and genealogies, globalism, trade, and the emergence of transnational corporations have altered imperial ideology. Yet mud continues to be deeply imbricated in more modern forms of imperial dominion.[29]

This book is divided into five chapters that address the relationship between these base substances and the process of empire building. Each chapter takes a base material—mud, mortar, ice, smallpox matter, paper, plasters, cloth—and traces the ways in which it is first celebrated as a useful, even enlightened, form of knowledge and then either discarded, forgotten, or rewritten as the fruit of British ingenuity. These materials function as technologies of empire: they are the things that make dominion possible, but only through a process that begins with sublimation and ends with abjection. Scientifically, sublimation is the chemical action of converting solid substance into vapor that then resolidifies upon cooling. Yet another meaning of this term refers to the transmutation into something higher, purer, or more sublime.[30] My use of sublimation engages both of these meanings; the substance of foreign techne transforms from solidity into an abstract truth claim; these alien forms of knowledge are transmuted by the alchemical paradigm into marvels. Here Alan Bewell's account of abjection in relation to Victorian disease is illuminating:

> Disgust was the assertion of a boundary both by individuals and the state, a means by which Europeans separated themselves from the larger world of disease. . . . On the other side stood the world created and rejected by abjection operating on a global scale, the excremental colonial world, whose prototype was India.[31]

Bewell relies heavily on Julia Kristeva's study of loathing as the primary drive defining the autonomous self. The body's rejection of filth structures a space in which it can imagine itself as free from the threats posed by the material world; as a paradigm of colonialism, particularly in the context of disease, abjection helps clarify the contradictory relations between the metropole and its colonial margins. Kristeva goes on to discuss the complexities of abjection, uncovering the primal and secondary processes of repression that structure this phenomenon of self-fashioning. Her account of the child's relation to the mother emphasizes the separation already enacted before the loathing that returns to complete this severance. In other words, she identifies a structural amnesia upon which the demarcated space separating the abject from subject is continually maintained.[32] Thus alien substance, at first hailed as valuable, once introjected as a material of empire building, becomes an object of fear, threat, and loathing, its initial value either scorned or forgotten. The structure of the British imperium, therefore, depends on an "enigmatic foundation."[33]

In chapter 1 I establish the relation between alchemy and description, unpacking an analogical paradigm that shapes the ways in which Britons "saw" India. In this chapter I identify the commercial and scientific excitement charted in letters from the early seventeenth century to the early nineteenth century. Drawing on the letters written by seventeenth-century ship captains and other East India Company factors, I flesh out how early anxieties about English commodities informed later propensities to grant Indian techne authority, as in the case of Helenus Scott. I argue that East India Company correspondence furnished important tropes for Alexander Pope to fulfill his poetic responsibilities to Queen Anne's court even while critiquing commodity culture. I look closely at *The Rape of the Lock* and *Windsor-Forest* to identify how alchemy informs Pope's ambivalent representation of England's ventures into the global marketplace.

Embedded in the language of alchemy is the sense of wonder that often saturated travel narratives promising, for example, so "inexhaustible a treasure of gold" as Defoe writes in *A New Voyage Around the World*.[34] Even if these narratives were not fictional, as in the case of the East India Company corpus, the descriptions of the magical or mystical features of foreign lands were often greatly exaggerated if not entirely fabricated. My point is not to determine the authenticity of travel accounts, a quixotic task at best. Rather, I am interested in the representational shift from India as a treasure trove of wonders in the seventeenth through the eighteenth centuries to India as a culturally impoverished and technologically backward site that prevailed from the nineteenth century to the present. I argue that, far from reflecting the inevitable rise of Western hegemony, a systematic nationalist amnesia produced by imperialist ideology structures this shift.

In chapter 2, I focus on mortar as a technology of empire whose history has been forgotten in the process of its abjection. That is, the fundamental material binding buildings together also sunders subject positions crucially apart. Using Isaac Pyke's treatise to the Royal Society, Jemima Kindersley's descriptions of the city of Madras in *Letters from the Island of Teneriffe*, William Langton's correspondence with the East India Company's Court of Directors, and Elihu Yale's tenure as governor of Madras as texts that address Indian mortar, I trace its history in Britain's first purchase in the Mughal Empire, Fort St. George. Although Madrasi mortar has a very clear material use in the construction of the city that eventually became known as White Town (and Pyke's description attests to its alchemical properties) by the time of Kindersley's visit, its history had been effaced. The stuff that held the Christian White Town together, borrowed from Black Town artisans, was now claimed as wholly English and secure from any possibility of contamination from its Indian origins.

In the next chapter I turn to ice. March 1775 proved to be an exciting time for Sir Robert Barker, stationed in India, traveling between Allahabad and Calcutta. Suffering as many British East India Company members did from the extreme heat and other forms of discomfort produced by seasonal monsoons, Barker wrote to the fellows of the Royal Society begging permission to offer his observations on the manufacture of ice in India, claiming that the Indians had discovered a method of making ice without the benefit of an appropriate climate. Barker's interest had as much to do with the availability of comfortable refreshment as it did with the prospects of climate control, of making India's alien weather English. Drawing on local knowledge, Barker's interest in Indian techne demonstrates the ways in which Enlightenment science and British colonialism were negotiated through the incorporation of "Asiatic" study. Barker's aims, however, were much more concerned with anglicizing the weather than with promoting Indian science. In much the same way that Samuel Johnson's dictionary provided a template for fixing the meaning of what it meant to be English, Barker's study capitalizes on fantasies of climate control that dominated the Little Ice Age. Here I also investigate the importance of ice as a technology of empire, as material, metaphor, and metonym of climate. By examining the sixteenth-century traveler, Bénigné Poissenot's *Nouvelle histories tragiques*, Samuel Johnson's *Preface* to *The Dictionary*, and Thoreau's journals to George Orwell's *Burmese Days*, I trace the various iterations of ice that radically shifted meteorological, agricultural, and social patterns in Europe.

Not all the substances and techne that East India Company men wrote about were new. When Robert Coult observed smallpox inoculation in Bengal, this practice had already journeyed from Constantinople to London in the letters that Lady Mary Wortley Montagu wrote to Sarah Chiswell some fifty years earlier. A great deal has been written on Montagu's interest in smallpox inoculation, from the letters describing the process to the inoculation of her own children. My own interest is in the letter writers Montagu dismisses as "common voyage-writers who are fond of speaking of what they don't know."[35] She acknowledges and admits to the peculiar trials of travel writers—"If we say nothing but what has been said before us, we are dull and we have observed nothing. If we tell anything new, we are laughed at as fabulous and romantic." However, it is abundantly clear that there are certain writers infinitely more reliable than others, and that class has a good deal to do with authoritative narration.[36] Steven Shapin summarizes the process of securing knowledge as a process of trust; that the "fabric of our social relations is made of knowledge—not just knowledge of other people, but also knowledge of what the world is like—and, similarly, that our knowledge of what the world is like draws on knowledge about other people—what they are like as sources of testimony."[37] Clearly, Montagu's social world was one inhabited by the aristocracy, and "common" travelers were not part of her circle of trust.

In addition to class, another psychosocial drive is at work in the production of knowledge. Smallpox inoculation was a deeply conflicted practice in England because of its foreign origins. Chapter four uncovers the xenophobia informing the scientific archive that, by the nineteenth century, had become so entrenched in English social relations that even after inoculation was rewritten as an English practice (vaccination) with an indigenous English history (William Jenner, James Phipp, Sarah Nelmes, Blossom the cow, and cowpox itself) and an English site (Gloucestershire), there was a deep social and cultural resistance to vaccination. Forgotten were the early eighteenth-century champions of this foreign techne, but even late eighteenth- and nineteenth-century advocates of inoculation were silenced, their scientific research locked in the annals of the *Philosophical Transactions* and certainly locked out of any part of Edward Jenner's "discoveries." The anti-vaccination movement of the latter half of the nineteenth century flourished, and their arguments against this practice invoked ideologies of nationalism, patriotism, and Christian communion. I use the correspondence between John Zebediah Holwell and Robert Coult with the Royal Society to complicate Edward Jenner's place on the medical roll of honor and to identify the ways in which texts as disparate as John Cleland's *Memoirs of a Woman of Pleasure* (1749) and Bram Stoker's *Dracula* (1897) reflect the contested ground of inoculation.

In chapter 5 I continue the investigation of Indian techne in the domestic, rural literary "heart" of England by turning to Jane Austen. Claudia Johnson enjoins us to pay attention to what lies hidden in plain sight; she argues that "our current recognition of Austen's artistic self-consciousness" is not due to the "discovery of any new information, but rather a disposition to pay attention to what has always been before us."[38] Johnson's remarks address Austen's clear interest "in matters as vulgar as commercial success," but I want to extend attention to other equally vulgar and equally visible matters in the *business* of authorship. In this chapter I enlist another alchemical transformation, one that reverses the sublimation of base matter. Using gender and class as ideological foci, I trace the ways in which intellectual work is demystified. Austen's novel *Emma* (1816), alludes to "offices for the sale of . . . human intellect" that advertise situations for governesses, a post that Austen clearly links to slavery and prostitution. Austen thus commodifies female intellect as intellectual property, and even while representing the delights of courtship this novel addresses with equal clarity the bleak and limited prospects available to genteel women without money. This form of commodification, however, was not the sole province of women. East India Company letter-writers commodified Indian intellectual labor every time they relayed a technological or epistemological "discovery" back to England. Without attributing Indian techne to any specific practitioner, Britons turned these forms of knowledge into property and harnessed Indian human intellect to serve British needs.

Austen's last novel, written in the early decades of the nineteenth century shortly before her death, records the labor of letters that, presumably, includes her own authorial work. It valorizes epistolary exchange without itself being epistolary. Epistolary labor assumes many forms in this novel: the representational work letter-writing performs, standing in for absent bodies and "bringing every body dearest to you always at hand," the labor of postal carriers who must decipher "hands" in order properly to deliver letters, and the "hands" themselves, formed by female tutelage.[39] In a more indirect way, Austen draws our attention to the abject materials of authorship and the tropes of labor embedded in paper, plasters, and pencils that are forgotten or discarded by more proper authors, but that are taken up, mystified and valorized by pretenders like Harriet Smith. *Emma*'s Highbury is adversely affected, as Maaja Stewart has argued, by the enormous amount of wealth brought in by British imperial culture, which had the effect of increasing "the poverty of the underclass and women of all classes."[40] Miss Bates and Jane Fairfax's palpable poverty reflects this imperial imprint.

But the British imperium is also marked by the labor of its colonizers. The East India Company servants who bring back foreign techne in their letters to the Royal Society offer methods of manufacturing paper to ease the burden of the high cost of cotton with which paper was made in Britain. Their letters note the sundry affairs of factories—inventory lists, negotiations, trade agreements, and other bureaucratic matters—and, in the case of early seventeenth-century correspondence, often the humiliations that they had to endure. For example, in 1613 captain Thomas Kerridge writes from Agra to Thomas Aldworth and the Council at Surat:

> I answered, the hat or anything else we -had was at the King's command, and entreated Cotwal [the Indian liaison] to tell the king I had important business with His Majesty, and desired audience. He went from me and stood before the king awhile but spake not to him and returning to me told me he durst not now move, the king being in conference with the Persian Ambassador, who stood before him, and I had my answer to repair to Marrobocan. . . . The next morrow went to the Marrobacan's house, attended there till noon and then word brought me, I could not speak with him, being with his women and came not forth this day. . . . The Kotwal sent for me and told me that the King expected I should send him my hat, I answered he should have it, and told him

I had been at the court and was denied entrance by the porters, he willed I should come in the afternoon and go unto him, he would give orders to the porters for my entrance at all times. But in the afternoon the King went forth to visit his father's sepulcher and returned into the city, but went to his tent pitched in an orchard some 2 coss off where he purposeth to make stay some 7 or 8 days.

Robert Travers mentions the cultural antipathy Persianate mughals had for "the 'hat-wearing' invaders, and their peculiar social customs and bodily habits." Thomas Kerridge's hat signifies his marginal status in this court: no matter how many times Kerridge offers his hat, he is refused an audience with the mughal. Unlike the rarified access to Turkish courts, which Lady Mary Wortley Montagu apparently enjoyed, Thomas Kerridge is without audience, subordinate to the Persian ambassador, the king's women, the king's father's tomb, and the king's pleasure party. Thus East India Company laborers, locked into correspondence with the councils of their various factories and with the Council of Directors in London, were able to enrich the English economy with commercial and epistemological success.[41]

The method I employ in this project focuses on relating historical fact to discursive event in relation to the archival recoveries of Indian techne. Thus the disquisition on mud—as a generalized phenomenon although I am well aware of the differences produced by localized semiotics—is an important consideration of the technology of colonialism because it deconstructs "territory" and returns it to its earthly material: the historical fact of "ground," later mystified as the discursive event of "colony." In the early modern period, I argue, travelers to India bumped up against this startling concurrence and, in order to make sense of what they saw, deployed a mode of analogical thinking. "Analogy," claims Ronald Schleifer, "stands between language (conceived as systematic semiotics) and the world (conceived in terms of more or less collaborative action). Analogy . . . traffics in constellations of wholeness."[42] Confronted with a series of events for which a *systematic* semiotics was not in place, early modern travelers to India read that world in "constellations of wholeness," allowing for the integration of foreign knowledge and foreign techne by *not* creating a hierarchy between principle and example, the familiar and the foreign, atemporal truth and historical fact. Elizabeth Ermarth argues that the securing—or "saving"—of truth rather than the saving of appearances was the global work of Enlightenment knowledge.[43] Replacing the "events" of analogy with the "facts" of analysis defined Enlightenment realism, but this particular strategy of mapping the phenomenological world ceased to make sense in places of extreme alterity. As seventeenth-century documents demonstrate—and later, as their literary counterparts represent—analogy was a means by which one could transform collaborative action, to use Schleifer's term, into a constellation of wholeness and thereby integrate foreign knowledge into a familiar systematic semiotics. The burden of this book is to uncover the ways in which a material object serves as a rhetorical connection, an understanding of analogy as the site where language encounters the world.[44] This site is also dependent on the alchemical sublimation of the material object into discourse: substance into truth claim. Connections between history and discourse, things and events are analogical ones, the explicit place where a signifying system encounters the material world.[45]

The Alchemy of Empire thus complicates the Eurocentric model of postcolonial inquiry, providing crucial supplements and, perhaps, correctives to the history of empire. It challenges the concept that "postcolonial" inquiry can only be conducted from the standpoint of the colony as the site of technological paucity. If I am guilty of reifying "England," "Europe," and "India" as geographically extant territories that are paired in structures of domination and subordination, I can only marshal Dipesh Chakrabarty's defense against similar charges and emphasize that "Europe works as a silent referent in historical knowledge" and remind readers that while "third-world historians feel a need to refer to works in European history; historians of Europe do not feel any need to reciprocate."[46] My book offers a paradigm of British anxiety and British technological—and epistemological—inferiority that was represented not only in the various reports from East India Company travelers, but also critically from powerful champions of British moral, commercial, and epistemological progress: from Alexander Pope's celebration of the Pax Britannica to George Orwell's bitter requiem to a dying British Raj.

The Alchemy of Empire

> For lo! The Board with Cups and Spoons is crown'd,
> The Berries crackle, and the Mill turns round.
> On shining Altars of *Japan* they raise
> The silver Lamp; the fiery Spirits blaze.
> From silver Spouts the grateful Liquors glide,
> While *China's* Earth receives the smoking Tyde.
>
> —ALEXANDER POPE, *The Rape of the Lock* (1712)

Alexander Pope's infamous coffee break of Canto III is inflected with just enough irony for his readers to wink and nod knowingly to one another, as they no doubt did at Button's coffeehouse after reading Joseph Addison's two-line review of the poem in *The Spectator.* Although coffee was good for the development of the literati and perhaps contributed to the transformation of the public sphere, Pope's account of this "grateful Liquor" in *The Rape of the Lock* is cautionary.[1] Coffee levels the political playing field by making all politicians "wise"; its vapors suggest new "stratagems" to the Baron to carry through with his nefarious plans while its fanned fumes distract the belle Belinda from her doom. Pope's admonition, however, extends beyond the parameters of poetic couplets, beyond Lord Petre's antics and Arabella Fermor's anger, beyond John Caryll's pacific pleas, and even beyond Bernard Lintot's pirated publication. Pope's warning even survives Pope himself who, not content to leave well enough alone, revised the original two-part poem of 1712 to the five canto epic beloved by readers from its final publication in 1717 to the present day. Enthusiasts of imperialist critique agree that the description of coffee and its attendant accoutrements corral spoils from the far corners of the earth and place them squarely in the middle of Hampton Court revelry. Just as Ellen Pollak complicated the figure of Belinda and Laura Brown and identified Belinda's dressing table, scattered with the booty of colonial conquest, as a showcase of British imperial power, so this scene of "rich Repast" similarly represents the commercial might of "Great Anna."[2] Admirers of Pope's irony would suggest that this inflated scenario expresses Pope's scorn, albeit veiled, for the whole business, citing the notorious "Great Anna" zeugma as proof positive of this disdain. As one of the most useful annotated editions of *The Rape of the Lock* suggests, "England began to secure its trading empire, and new materials and luxuries poured in from around the world."[3] More recently, Suvir Kaul argues that poets of the long eighteenth century were conscious of the unique capacities poetry provided to articulate what he terms a "vocabulary of nation," and Pope certainly stands as one of the more powerful voices of this work in progress.[4] Thus, the groundbreaking readings of Pope Laura Brown and Ellen Pollak offered in 1985, readings that heralded the inception of studies of nation anchored in the acknowledgment of imperialism had become, a sparse fifteen years later, an interpretative assumption.[5]

But what if Pope's caution responds to something quite different from either the careless consumption of an indulgent aristocracy or the deliberate display of British imperial power? The key to this lock is in "China's earth." Obviously, Pope's fanciful figure refers to the china cups on the "Board," and the meal spread out to entertain Belinda and her cohort is enriched not only by the coffee—a relatively new commodity—but also by the china and the silver service, all of which is quite fitting for an afternoon of royal entertainment. These lines enunciate sotto voce a shift of monarchical power. "Great Anna's" authority, put into question by two insistent zeugmas that conflate her effective counsel with tea and her majesty with an "Indian screen," is figuratively dethroned. Pope replaces her crown with china teacups and silver spoons. Far from being the relatively simple components of a relatively domestic scene, these cups and spoons are metonyms of other sovereign powers: on the one hand, the Ming and Qing dynasties controlling the production and traffic of silks and china, and, on the other, the Spanish empire of the New World controlling its putatively inexhaustible supply of bullion.[6] Sandwiched between the powerful imperial terminuses of "China and Peru" is Britain that, at the time of *The Rape of the Lock*'s first appearance in 1712, was still engaged in Queen Anne's War, its eventual victory not yet secured.[7] Pope, himself on the brink of establishing a prominent poetic and political voice, was similarly poised between two versions of the poem. One version, characterized by "the satirical delicacy of its self-sufficient world,"[8] and the other that tackles "the power of mercantile and colonial modernity to reshape individuals and nations,"[9] between which he revised, at Queen Anne's behest, his 1704 pastoral, to celebrate the Pax Britannica and produce *Windsor-Forest*.

The "Cups and Spoons" enter the poem as discrete commodities but quickly transmute to the material with which they are made. Spoons become silver and cups return to "China's earth." In a poem where a "Goose-pye" speaks and a "Pipkin" walks, this transformation shouldn't be that surprising. But unlike the sly, sexual innuendoes that playfully echo in the Cave of Spleen, these cups and spoons represent something far more anxious. The silver spout empties a "smoking Tyde" into "China's earth." Dismantling the china cup and rendering it back to its original substance, earth, has the curious effect of robbing the cup of its function: as "earth," the cup absorbs the tide and its silver rather than acting as a mere receptacle. Putting a little pressure on the word "Tyde" for a moment, we can read another tide, born of the proud "Silver Thames," upon which ships laden with silver bullion sail to China to exchange this precious political cargo for the trivial commodities that litter Belinda's dressing table and delight the gossips at Hampton Court Palace. André Gunder Frank's study of global economy provides interesting statistics on the amount of silver flowing into continental Asia. Frank claims:

> over the two and a half centuries up to 1800, China ultimately received nearly 48,000 tons of silver from Europe and Japan, plus perhaps another 10,000 tons or even more via Manila, as well as other silver produced in continental Southeast and Central Asia and in China itself. That would add up to some 60,0000 tons of silver for China or perhaps half the world's tallied production of about 120,000 tons after 1600 or 137,000 tons since 1545.[10]

These numbers endow Pope's figurative representation of China's "earth" with impressive materiality, even to the support offered by Japan's "shining Altars," gilded, perhaps, with 10,000 tons of silver it was able to export to China. Kenneth Pomeranz's influential study, *The Great Divergence*, similarly attests to Chinese interest in silver:

> When Westerners did arrive, carrying silver from the richest mines ever discovered (Latin America produced roughly 85 per cent of the world's silver between 1500 and 1800), they found that sending this silver to China (whether directly or through intermediaries) yielded large and very reliable arbitrage profits—profits so large that there was no good reason for profit-maximizing merchants to send much of anything else.[11]

Thus the capacity of Chinese mud to incorporate most of the world's liquidity and retain the power commanded by bullion poses a very real threat to a culture like Britain, newly addicted to Eastern commodities but, unlike Spain, lacking a reliable source of bullion. The glittering spoils veneering *The Rape of the Lock* may not unilaterally represent Britain's imperial power. Whirring in the background of chat and "all that" is the inexorable "Mill" that slowly grinds distracting and destructive vapors for "Heroes" and "Nymphs" at play, and provides a machine of eternal damnation for errant sylphs ("The giddy Motion of the whirling Mill").

Interestingly, Pope deploys a reversed alchemical trajectory to give heft to his cautionary tale. Rendering china back to its original base substance may seem to rob it of its immense value and recast it as a trivial item, part of a long conceit of silly pleasures indulged by coquettes and beaux. Ultimately, however, such a rendition demonstrates British helplessness in the face of Chinese power. As a mock-heroic epic, we are meant to take this caution with a grain of saltpeter, but Pope's insistence on trivia and the triviality of poetic action suggests that far more is at stake. Even *Windsor-Forest* expresses ambivalence toward the putative pax Queen Anne had created with the signing of the Treaty of Utrecht. Under Granville's command the timber, for example, from which ships including East Indiamen were made leaves the forest forsaken of half its trees. Pope ends the poem somewhat melancholically with a wistful reminder that his "careless Days" were much more profitably spent singing "Sylvan Strains" to "list'ning Swains" than celebrating nautical power.

"The Same Wind"

I now want to identify the ways in which Pope harnesses the movement between substance and commodity to articulate his critique. Kaul points out that Pope's focus on the trivial stabilizes and domesticates the "historically transformative force of imported commodities."[12] But equally important to the narrative of nation making is the acknowledgment of other sovereigns and other empires which Pope addresses in his early eighteenth-century anthems.

By mid-century, however, this circumspect voice was less evident, and no text has been more drafted into postcolonial service than Samuel Johnson's *Rasselas* (1759), particularly the moment when the protagonist asks of his teacher, Imlac: "By what means are the Europeans thus powerful? Or why, since they can easily visit Asia and Africa for trade and conquest, cannot the Asiatics and Africans invade their coasts, plant colonies in their ports, and give laws to their natural princes? The same wind that carries them back would bring us thither?"[13] Imlac replies with an object lesson in ideology: "They are more powerful than we because they are wiser; knowledge will always predominate over ignorance . . . but why their knowledge is more than ours, I know not what reason can be given, but the unsearchable will of the Supreme Being."[14] Imlac's invocation of this "will" effectively dismisses Rasselas's reasonable question. The idea that rational thought was solely the province of Enlightened Europe had assumed hegemonic status by the mid–eighteenth century and Johnson is only ventriloquizing what seems a self-evident reason for European power. Since that time, the reasons that explain Western superiority in the eighteenth century have changed but not the assumption of that superiority. Reason for Johnson unambiguously endows the "great number of the northern and western nations of Europe . . . which are now in possession of all power and all knowledge," while postcolonial readers see the impulse of Enlightenment reason in the very the logic of Rasselas's question.[15] That is, Reason has become the reason for Western domination.

Historians and anthropologists interested in the history of empire have invoked Johnson's sentiment, using this moment to illustrate variously the ways in which European hegemony operated in the eighteenth century. Jack Goody, for example, uses this exchange to historicize European ethnocentrism: comparing this moment in Johnson to another one in Shakespeare's *Richard II*, he demonstrates reason to be the discourse that makes Occidental superiority possible.[16] The same Johnsonian moment functions differently for Niall Ferguson. In his *Empire: The Rise and Demise of the British World Order and the Lessons for Global Power*, Rasselas's question appears as a self-congratulatory epigraph to the chapter "Why Britain."[17] I want to question the assumption of European Enlightenment as it was developed in seventeenth- and eighteenth-century Europe, and as it has been extended in academic and cultural work since, particularly the notion that its Reason belonged uniquely to the West. In fact, both Johnson's assumptions about the unique intellectual attainments of the European Enlightenment and today's critique of that assumption are consequences of European colonial triumphs. Neither recognizes the colonial sources of what we take to be modern reason.

Ferguson's account of Western history uncovers the ways in which scientific knowledge, mechanical expertise, and technological development contributed to the increase of material wealth in the West via new forms of commerce. For him, these developments reflect a straightforward progress based on the application of newly discovered models of thought manifested in mathematics, chemistry, economics, and so on. The first beneficiaries of these technologies lost sight of their origin and saw them as the product of a divine gift of Reason. Ferguson clearly reveals the delusions at the heart of Enlightenment privilege, yet his work remains bound to a narrowly Eurocentric model. If, however, we attend to other more global histories, then we may be able to see that there were in fact competing technologies of empire.[18] Although historians have effectively exposed the ideological Orientalism in Johnson's tale, the rhetoric of that tale reveals an even more compelling epistemological history. When Johnson invokes the power of the "same wind" endowed with the capacity to blow ships of all nations into lucrative ports, when he sees that the potential to establish commercial, technological, and epistemological connections equally among the nations of Asia and Africa and Europe and the east, he raises the possibility only to sweep it away. Without the divine will of an unsearchable intelligence, the east can no more challenge western dominance than a woman can preach. But such a dismissal reveals something about the fears that plagued Johnson's contemporaries.[19]

Uncovering the early history of the shift toward Western scientific hegemony has been traditionally understood as a progress narrative, a chronological succession of accumulated knowledge culminating in a grand imperial telos.[20] Reading disparate letters, journals, and diaries, written by various early modern merchants, soldiers, doctors, scientists, and scholars who traveled in the East may help uncover ways to complicate this progression. More important, such recovery may identify different, non-Eurocentric discourses that the British used to make sense of new spaces they encountered in the seventeenth and eighteenth centuries.

The ideological lesson Imlac teaches Rasselas is one that has to be rhetorically structured. Even if Rasselas were to convince Imlac of the various capacities for restoring epistemological agency to Asiatic and African nations, the trope doesn't work. The "same wind," in fact, does not blow in any static fashion. That "same wind" is subject to different currents, different directions, and harnessing the "wind that carries them back" to bring "us thither" would involve different ways of charting courses, sailing ships, forms of currency, forms of knowledge—in short, a different techne. Rasselas's question and Imlac's answer can only be understood through Johnson's assumptions about commerce, technology, and epistemology: most telling, the larger assumption about the reciprocity of these cultural practices. Johnson therefore articulates his culture's fears and anxieties about a different techne. The putative ease with which these Europeans navigate the seas to "visit" Asiatic and African nations masks the necessity for trade and conquest. That is, for Britons seeking to supplement their meager island resources with trade—for spices, silks, silver, and such—these voyages are necessary. For those Asiatic and African nations, however, whence these luxuries originate, there is no need for sea-faring exploration because there is no similar need for them. The same structure of desire is not in place; there is no transcendent sense of reciprocity necessarily governing the asymmetrical relationships between Europe, Asia, and Africa. These nations are themselves self-sufficient, autonomous, and free standing. The possibility of conquering European nations cannot present itself as a possibility to them not because of Europe's epistemological mastery but because trade implies an economic exchange that seems not to have been historically in place.[21] Contrary to most (European) estimations of the limitless capacities for their trade with Asiatic and African nations, Western Europe had little to offer in the way of commodities. Especially in terms of manufactured goods, Europe and England had little effect on Asian markets and on the fashion sensibility of their courts. By comparison, Asian goods had a profound influence on European markets, shaping the consumerist practices and tastes at a very rapid rate.[22]

Pomeranz argues that the historical accounts of the accepted divergence between the development of European overseas commercial practices and those of the rest of the old world, particularly in pre–nineteenth-century Asia may not be as profound as it has been otherwise represented. The formation of joint-stock companies and partnerships that Western Europeans favored for the management of overseas trade facilitated, for them, the management of enormous cargoes and trading expeditions that would, according to their calculations, be far too large for a single investor. "By contrast," Pomeranz writes, "a Chinese junk trading in Southeast Asian waters typically carried the stock of many different merchants, and those merchants or their agents were onboard serving as crewmembers and receiving cargo space in lieu of wages." Such a system, he continues,

> made perfect sense in light of the monsoon winds on these routes. Since one could not return home until the winds reversed themselves, one could not drastically reduce the amount of time spent in port(s). A group of land-based entrepreneurs who used all the cargo space themselves or rented in for cash would have found themselves paying enormous wage bills to support professional sailors during the long stays on shore. It made more sense to make many shorter stops and have crewmembers who could and did occupy themselves in buying and selling cargo in each port.[23]

The structure of Asian commerce overseas, at least as long as ships relied on wind for their power, is one that defies Johnsonian logic. That logic imagines the same pattern of winds across different commercial terrains. Asian systems, by contrast, were acutely aware of wind patterns and adjusted their commercial activity with those differences in mind. Certainly their trade was not with Europeans or the British but with other powerful sovereigns of the Indian Ocean. In other words, while the "same wind" blew profitable commodities into the English marketplace, the reverse was not always true. If there was no need on the part of African and Asiatic nations to exchange their materials and technologies with the British, for whom African and Asiatic material and techne were crucial to national and cultural survival, then the only means for the British to secure supply routes was through conquest.[24]

Given British interest in cartography and taxonomy in the eighteenth century, it seems that Johnson would have produced a more efficient answer to Rasselas's question than Imlac's vague reply. It seems that Johnson would have attended more closely to the specific material reasons for European commercial, technological, and epistemological dominance. I am interested in what is swept under the auspices of the "unsearchable will": the numerous and conflicting economic, political, and historical problems with "Asiatic" trade that often troubled aesthetic accounts of British national integrity. Eighteenth-century Britain was deeply invested in searching, naming, understanding, collecting, and, most important, seeking purchase in the world around them. The "Supreme Being," variously named but most often invoked as a form of Providence, seems curiously uncharted in Johnson's representation, an articulation of the failures of taxonomic or cartographic language to represent the misfortunes of British exploration or the breakdown of commercial exchange.

Concrete issues with "Asiatic" trade threatened the British presence in a global marketplace and suggest that British goods and British technology did not necessarily measure up to Asiatic standards. The implications for the English, deeply addicted to new commodities flowing into their markets, are conflicted and conflicting, as poems like *The Rape of the Lock* indicate.[25] The ideological lessons embedded in *Rasselas* may reflect an English sense of global trade, but, as Johnson's rhetoric also represents, the logic of this sense is flawed. A closer look at the correspondence between British East India Company members posted to Mughal India and the Royal Society in London uncovers a profoundly different "Oriental tale."

As late as the 1790s, Helenus Scott—officer, gentleman, doctor, adventurer, observer, and author—wrote quite frequently to Joseph Banks, head of the Royal Society, with his "discoveries." These letters were often punctuated with tremendous excitement, and though they were tethered to his commitment to the scientific decorum established by Enlightenment philosophy—the most insistent of which was the relationship between class and testimony—he was nevertheless quite convinced by the power of Indian techne. "You will think the paper for replacing noses on those who have lost them an extraordinary one. I hope to send you by the later ships some of the Indian cement [*Caute*] for uniting animal parts," Scott writes to Banks in 1794.[26] This cement did not materialized in London—or, if it did, little mention was made of it by Royal Society members—but the paper Scott wrote contains specific details of this act of unification. Quoting from James Findlay, the surgeon of the Company's residency in Poona, Scott describes the method and substance of Indian rhinoplasty:

> Through Sir Charles Mallet's obliging influence Mr. Cruso & I were permitted to see the operation performed on the 26 ultimo by a man of the Koomar Cast (a class of Hindoos chiefly employed in making common earthen ware of this Country) who, with an old Razor borrowed on the occasion, dissected with much composure a portion of the frontal Integuments from the pericranium of the Patient & grafted it, a new operation to us in Surgery, on the stump of the original nose. He then retained it by a Cement without the use of stitches, sticking plaster, or bandages. The Patient is at present in good health & high spirits. An adhesion has taken place seemingly in every part; when it is perfects & cicatrized I shall give you a particular History of the operation of subsequent treatment.[27]

Unlike smallpox inoculation, which was only administered by a sub-caste of Brahmins called Vaidhyas, this medical practice was on par with making the humble earthen wares of everyday domestic use. But the structure of medical practice was far more complex than Scott's description implies; the "composure" with which this "man of the Koomar Cast" performs the operation is born of centuries of cooperative epistemological exchanges. Zaheer Baber describes the social organization of medieval Indian medical practice:

A specific hymn from the *Rgveda* indicates that in terms of social status, medical practitioners and carriers of this knowledge were placed in the middle of a threefold list of skilled professionals that included carpenters (*taksan*), healers (*bhisaj*), and priests (*Brahman*). Thus, physicians and healers combined the craftsmanship of a carpenter with the intellectual acumen of the priest. Like the learned priests, the healers commanded esoteric knowledge, and like the uneducated but skilled carpenters, they "repaired" the injured or broken human body. Within the social hierarchy of the early Vedic period, they were respectd and even praised in the *Rgvedic* hymns for the healing services they performed.[28]

The caste that putatively attended to these operations was primarily constituted by artisans, but to the Europeans they performed feats of surgical wizardry hardly to be believed. Scott claims "such a practice was not known among the Europeans," noting that the "*Cement*, by which the old and new parts are kept together till they unite, appears to be a desideratum in our Surgery,"[29] and concludes his letter "beg[ging] leave to repeat that two of the Company's surgeons, M. Findlay & M. Cruso both men of eminence in their profession have actually seen the operation performed & the Sepoys who are mentioned above are known to all the gentlemen at the Residency at Poona" in an attempt to stave off disbelievers.[30] Scott effectively transforms artisans of earth into medical masters.

Strangely, the *caute* that excited Scott's interest did appear in London, but not as the desideratum he mentioned in his correspondence with the Royal Society. The word Scott employed was probably a mispronunciation of the Hindi *Katha* for the medical astringent catechu, a substance used in *Ayurvedic* practice. This astringent, extracted from the acacia catechu plant by boiling the wood and evaporating the resulting brew, was also used as a tanning and dyeing agent for wool, silk, and cotton. It was brought to Europe in the seventeenth century, probably by Portuguese and Dutch merchants, and because it looked like earth, it was commonly known as *terra japonica* (japan earth).[31] A strange relation exists between images of earth and descriptions of Hindu techne: potters perform complicated rhinoplasty while deploying, in turn, an exotic and orientalized "earth" to magically adhere body parts together.

Somewhat less dramatically, Scott sends "a piece of cinnabar of this country which is made in masses of 100 lb. weight," having "frequently tried to make cinnabar by the methods recommended in Europe," but having "not been able to procure any so far, as the Indian at one operation."[32] Scott has also been "for several years attentive to the methods used by the natives of this country for dyeing their cotton cloths," and thinks that he has "discovered the singular circumstance" that gives "permanency to the colour which is so much admired [by Europeans]." Yet he is "unable to give any theory of the operation of the chief Substance they use" except to suggest that "a cloth is wetted with an infusion of it [the mysterious substance] and a solution of alum," the result of which renders "cloth and colouring . . . ever afterwards inseparable . . . If this appears to you a matter of consequence as the cotton manufactures are now in so flourishing a condition in England I shall at some future period communicate more particularly their method to you." Scott's query at the conclusion of this letter is typical of his relationship to Joseph Banks and the Royal Society in general: he wishes for their sanction, to be one of the soldiers of England's scientific and epistemological fortune and to contribute to the continued success of English manufacture. In supporting English techne, Scott implicitly critiques the "flourishing" condition of English cotton mills. The "chief Substance" that renders such "permanency" of color to cloth is plainly absent from English methods of dyeing. Although Scott sees certain clear results—textiles saturated with brilliant and fast color—the "*chief* Substance" of the method remains mystified, perhaps because Scott could not see or obtain details of the technique. Eluding his observational eye is any kind of documented knowledge he can recognize; analytical reason fails to make this "chief Substance" visible. Scott expresses his aggravation, writing "their knowledge of the arts is never communicated by writing nor printing nor their experience reduced to general laws by theory, the difficulty of gaining information is again increased,"[33] and therefore cedes his authority to Indian artisans by mystifying their operations in his communiqués to London.[34]

Helenus Scott transforms base material to wondrous substance and uses this transformation as a model for their accounts to the Royal Society. Enjoining Joseph Banks to analyze a vegetable astringent he sent, he writes: "in fixing some colours it has hidden powers."[35] These accounts constitute proper specimens of the "Indian arts" that encourage Scott to find himself "sufficiently repaid if [the Royal Society] thinks [he] has contributed anything to the interests of science." Furthermore, "should anything [he] transmits appear worthy of printing, [he] can have no objection to it."[36] It seems relevant that these practices, the descriptions of which range over the eighteenth century, demonstrate the East India Company's enduring interest in alien techne. It is also relevant that they are often discovered to be superior to their European counterparts especially given the increased authority science commands in the West. Nineteenth-century views of Indian science and technology were sustained by the historically transformative power of the industrial revolution that changed the shape of the marketplace in England and solidified colonial ideologies. Thus the condescending British views of Indian science and technology: they were "versions" of "science," increasingly understood as medieval, worthless, and clear signs of Occidental epistemological superiority.[37]

This was the assumption popularly received and popularly disseminated. However, a cursory glance at the *Philosophical Transactions* from about 1680 through the 1790s seems to disprove this idea. The reports that were published by the Royal Society manifestly indicate the anxieties expressed and felt by British entrepreneurs: their goods and their technologies did not measure up to the commodities and techne available in Asia long before Britons started their maritime adventures. Helenus Scott sent a long treatise on an early form of carbon steel manufactured in the south of India called "wootz," whose properties were examined and tested by the Royal Society. The report, published in the *Philosophical Transactions*, soundly endorses this substance, and recommends using the Indian methodology to supplement British steel manufacture.

As early as the seventeenth century, East India Company factors wrote frustrated letters to the office in London expressing the disdain with which Indian Mughals and Portuguese rivals alike looked on the goods these newly arrived merchants brought to trade. Nicholas Downton, for example, writes in November 1614: "It seemeth to me the ill sales of cloth in India put Mr. Aldworthe into an extraordinary desire by Inquisition to seek out a better place in regard of their cloths yet remaining on their hands, as for such as he feared were to come by the next shipping, and the next after that, before advice can be sent home to forbear."[38]

Thomas Kerridge writes earlier that year: "The Viceroy of Goa in a letter lately written to this King wrote very basely of our nation, terming us thieves, disturbers of states and a people not to be permitted in a commonwealth, and that if the king received us they would never have peace with him."[39] But these were early British traders, unsure of their footing with powerful sultans and European rivals, willing to accommodate native desires and whims in order to gain some small purchase in an enormously lucrative market.

What Royal Society publications demonstrate is that British acquiescence to local rulers and local technologies continued throughout the eighteenth century, although under a different guise. Stories of British mercantile and cultural encounters with India were disseminated in England as histories of conquest and control; East India Company correspondence, however, represents epistemological encounters that tell a very different tale.[40] These writers, pressed to describe marvels from the old and new worlds, invented a language of mystification to represent modern scientific "truths." What did this mean?

K.N. Chaudhuri usefully deconstructs Eurocentric notions of empirical reason. Addressing the complicated relation between chronology and epistemology, Chaudhuri argues that the "implied assumption that the period from the seventh to the eighteenth century constitutes a chronological unit for Asia as a whole needs to be examined in detail and its rationale excavated." Early in the European eighteenth century, Chaudhuri writes, an "Umayyad or Abbasid Muslim would have found it decidedly strange, if not actually beyond the limits of comprehension" that the Rumis and Franks were posing a threat "to the Muslim right of free navigation in the Indian Ocean." Such a person would have, above all, "found it incredible that the same Europeans now held the key to collective wealth and material well-being through their knowledge of the sciences and practical arts."[41] For him, the historical shifts in commercial power simply would not have registered as serious competition and certainly would not have been perceived as the inevitable result of European progress.

The substances that constituted such marvels in the eyes of the servants of the British East India Company had historical significances remarkably different from what later became the domain of the intuitive. For example, later in the nineteenth century, some of these substances dropped out of scientific favor and disappeared from the British scientific consciousness, which implied a substantial shift in the structure of epistemological authority. Nevertheless, these substances remain stubbornly lodged within the archives of scientific discovery. Unearthed, they produce a paradigm shift (to borrow a term from Thomas Kuhn) in the history of British imperialism.[42]

The reports written to the East India Company and the Royal Society extend over the course of the eighteenth century and demonstrate that European scientific hegemony was not as solidified as it is represented in later histories of European hegemony. In particular, contemporary historical stories about globalization and empire written by Niall Ferguson and his ilk foreclose with stunning regularity the clearly powerful positions Asian empires occupied, not only in the sixteenth and seventeenth centuries but throughout the eighteenth century. Even early into the nineteenth century, such disbelief in European commercial, technological, geographical, and, arguably, scientific mastery persisted. Mehmed Said Halet Efendi, the Turkish ambassador to France from 1803 to 1806, "believed that the European economic competition could be effectively broken merely by setting up five factories for snuff, paper, crystal, cloth, and porcelain, as well as schools for language and geography."[43] Halet's suggestion implies that these institutions of manufacture and education would be sufficient to quell European interests in the Muslim market and reorient European ideas about power and agency to acknowledge Ottoman hegemony. Halet's comments challenge historical and contemporary Eurocentric ideas about the primitiveness of Asian science and technology. Interestingly, the reason Halet seems to arrive at these conclusions is not necessarily the analytic mode prized by Enlightenment empiricists. Rather, he seems to deploy an analogical understanding of the problem: we approximate their putative technological and epistemological expertise and by example make them understand their shortcomings. Obviously, these five factories were not enough to stave off European intervention in the Asian marketplace, certainly not in the nineteenth century. Yet what this somewhat fanciful notion suggests is how each empire imagined a hegemonic position based on its sustained belief in the reciprocity of perspective. Earlier in the eighteenth century, British East

India Company discoveries and representation of fantastic substances deployed a similar paradigm of reason, but one that led them to understand the fabulous properties of Indian substances and, by comparison, the relative paucity of their own resources.

The relationship between exotic substance and sublimation may tell us a great deal about the truth claims these British East India Company employees made in order to contribute to the interests of Western science. Their efforts to translate an ancient and (to them) mysterious knowledge of the East into terms that approach a kind of scientific sublime contribute to a discourse of invention and usefulness that both informs and resists traditional conceptions of Western modernity.[44] Such efforts may also attest to the failure of Enlightenment empirical reason, such as the form employed by Rasselas and rejected by Imlac, to account for the one-way trade routes European merchants represented themselves as navigating. What this motley collection of letter writers may have discovered is that the reigning paradigm of Enlightenment reason may not always have been the most useful or even the most immediately accessible way to make sense of the world they observed outside of Europe.

Alchemy and Reason

Alchemy provided a model of reason that did make possible the transformation of material substance into a scientific sublime. The model these readers had in mind was not the popular belief of the mystified transfer of lead or any base material into gold. It was a model that emphasized the importance of practical knowledge.[45] Pamela Smith's *The Business of Alchemy* outlines in enlightening detail the adventures and history of a particular alchemist, Joachim Becher, who self-consciously fashioned himself an advocate of practical knowledge, or praxis: the mutable affairs that could be directed and intervened by humans. Such a paradigm of material knowledge differed from the earlier medieval model that privileged a circumscribed understanding of epistemology that could not be shaped or directed by anyone outside of the church. Mathematics, a discipline now almost entirely connected with abstract thought, was in the seventeenth century mostly identified with trade.[46] An English contemporary of Becher's, John Morris, offers this account of mathematics: "For mathematics . . . were scarce looked upon as *Academical* Studies, but rather *Mechanical*: as the business of *Traders, Merchants, Seamen, Carpenters, Surveyors* of Land or the like; and perhaps some *Almanack-makers in London*" (1690).[47] Originating from material sources, Newton and Leibniz helped mathematics become the language of disembodied knowledge and a scientific discourse associated with "truth."[48] Mathematics sublimated from a material base to a pure abstraction of "truth."[49]

Alchemy functions as the historical link between the divine and the material, the methodology that seeks to make truth material and valuable. Helenus Scott's interest in the Indian sublimation of cinnabar (the red mercury distilled from this substance was one of the more important materials used in alchemical practice), expressed late in the eighteenth century, suggests that the belief in alchemy was still persistent. The Alexandrian practice of this form of transmutation, traced back to the Babylonian tablets of the thirteenth century BC (which, according to Debus, also "describe the production of 'silver' from a copper-bronze mixture"), was transmitted to Europe through Spain by the Arabs who also transmitted the numerical system that became associated with alchemy.[50] Enlightenment science provided a methodology that contributed to the later hegemony of European epistemology. This science was based on alchemy and was not the exclusive provenance of Europe; rather, it was produced through an epistemological exchange between Asia, North Africa, and Europe. These forms of knowledge became the discourse of progressive science, their origins forgotten or erased until an encounter with alterity made their resuscitation necessary.

Such practical models of knowledge that depend on human observation need to be sanctioned by the same authority granted to the divine: the bodiless understanding of truth that is founded, in the motto of the Royal Society, *nullius in verba*. Steven Shapin's *A Social History of Truth* identifies the relationship between knowledge and trust. He suggests that if "the natural world was a great treasure trove of hitherto unimagined marvels and singularities, the legitimate scientific practitioner was by no means obliged to credit *all* pertinent knowledge-claims. Marvels indubitably existed, but they had to be authenticated as such: *this* marvel-report had to be verified."[51]

English natural philosophers and historians schooled in gentility and civility provided the necessary authority for credit to be given to others' observations. Sir Charles Wolseley, for example, writes that

> to deny credit to testimony is to deny ourselves the benefit of any part of the World, or anything done in any part of the World, at any time in the World; but just what we our *selves saw* in the time and places therein we lived. No one Age can be of any use to another, but his Credit . . . For, the same reason, that will make a Man not to believe others, will be as good to them, not to believe him: and so, all Mankind must *live upon their eye-sight.*[52]

Empirical reasoning, therefore, was absolutely contingent on the idea of crediting the testimony of others. For Helenus Scott and for other East India Company merchants, writing from a strange and mystified land, that meant granting them the privilege that his observations and reports followed a strict scientific methodology. The fact that many of these reports found their way into the *Philosophical Transactions* suggests that representations of new technologies and methodologies were given not only the benefit of the doubt but the benefit of the supplement: using these new substances and technologies to secure European epistemology.

Even if these writers to the Royal Society employed modern and "progressive" analytical modes of reason—ones that reduced the visible phenomenological world to its disparate, reductive, constant, and, according to an empirical model of reasoning, immutable parts—their encounters with foreign and exotic substances could not register as a seamless part of European or British taxonomy. Rather, these writers required the use of another epistemological paradigm: one that privileged an analogical versus analytical form of reason in order to make sense of the world they thought they saw around them. It was not a convenient return to an "earlier" model, but rather an implicit understanding that one discursive paradigm was contingent upon the other that made possible any kind of articulation in these encounter narratives.

Empirical reason dominated scientific practice and epistemological decorum in eighteenth-century Europe. Other forms of testimony, however, were needed for the members of the British East India Company, anxious to serve the cultural progress of their nation, to account for unfamiliar marvels existing outside of their taxonomic range. These intrepid travelers didn't merely fold foreign oddities into their extant taxonomies. Their reports demonstrate an epistemological problem that is far more complicated. Early English encounters with strange substances drew upon a complete dependence on analogical reasoning that granted authority to Indian epistemology. By the nineteenth century, this form of reason had been appropriated through colonialism and the hegemony of analysis and duly devalued. Thus, room was made for the answer Imlac gives to Rasselas's question, an answer that reflects domestic complacency because British economic, cultural, and technological encounters with the other had already been altered by Indian techne. Helenus Scott and his company demonstrate a different idea of encounter; they were convinced throughout the eighteenth century and into the nineteenth, that "luxuries" emanating from exotic and mysterious substances and practices were the result of the "Asiatic's principle study," an epistemology, technology, and aesthetic to which they had no access save what they saw themselves.

One of the striking characteristics about the discoveries these writers made is the insistent mystification of substance. Transforming base material into something with marvelous, wondrous, or magical properties could be, at the very least, something that would prove fantastically profitable for the East India Company, the Royal Society, the Ship Builders of Leadenhall Street, or for the new Indian spaces Britons were beginning to inhabit. Possibilities of profit took shape as new forms of commerce, new scientific discoveries, and new ways of waterproofing ships, new techniques and formulas for building materials. The sublimation of mundane material into something with infinitely more value follows the alchemical pattern of turning the material into the sublime: something that at once signifies embodied and disembodied value.

The seventeenth through mid-eighteenth centuries were tumultuous times for the British East India Company. Without a solid purchase in India's profitable marketplace, Britain found itself in the uncomfortable position of having to account for its own economic failures. Nicholas Downton, for example, early in the seventeenth century exhorts the Court of Directors to stop sending useless woolens to sell in Surat. William Edwards informs the Company that the "small commodities" they have chosen to trade with the Mughal courts on the Malabar Coast have little use other than as "presents," and urges them to take better care of gifts that were much more valuable for his purposes because they marked a possible ingress into a future market in other commodities. He writes:

> All the small commodities which were sent in these ships, as looking-glasses, comb cases, knives, pictures, fowling pieces, Muscovy hides, and such like, serve only for presents, but will not *sell* for any price. (emphasis added)

Whereas:

> if it please you to send by your next ships . . . an English coach and coachman, to bring their horses to that labour, it would be very acceptable with the king; and to send some curled water spaniel of the greatest size, with a bloodhound or two, they would be very welcome, for *they will hardly be persuaded that they can* be *taught to fetch and find things lost.* The mastiffs that came along in these ships are all dead except one, whereof we are very chary, for that I understand it will be very acceptable with the king.[53]

Underscoring the need to be able to sell English goods and noting the things that pique Mughal interest—coaches, dogs, and the various forms of labor that both can perform—Downton acknowledges the precarious place of British merchants. Downton concurs with the value English fauna (whether real or representational) and other trinkets have to the Indian Mughal, writing in his "Particulars desired from next ships from England to Surat for Great Mojore" (1614): "crooked swords, all manner of toys for women, pictures in cloth, not in wood, any figures of beasts, birds, or other similes made of glass, of hard plaster, of silver, brass, wood, iron, stone, mastiffs, greyhounds, spaniels and little dogs, three of each."

He notes that "Figures of divers beasts and dogs in stone and plaster I have seen come from Freinckford [Frankfurt]. I think at Amsterdam may enough be had" and that "Dogs hard to be carried."[54] Later letters exhort the Company directors to take better care of the transport of live animals, noting that the death of dogs could be easily prevented by making sure they were watered properly. The wild calculations made by British merchants as to what constituted commodities proper to Asian trade—woolen cloths, small trinkets—vastly underestimates the economical, commercial, and technological mastery wielded by Indian sovereigns as well as their political power in the marketplaces of the Indian Ocean. Far from commanding any kind of epistemological authority, merchants like Downton and Williams quickly discovered that they were in the awkward position of procuring commodities that signified the existence of an English cultural landscape to vast and powerful courts of disbelievers. At the very least, these querulous requests suggest that Company merchants were primarily at the mercy of the Indian and other European rulers (like the Portuguese) alike, and it was in the East India Company's best interests to reconceive the commodities that would serve their own power vis-à-vis their non-English rivals. English merchants could not sell anything in India except gold, silver, and saltpeter for gunpowder.[55] Indian markets ascribed to different systems of value that the British merchants had to learn if they were going to be at all successful in their trade. But even later in the eighteenth century, when domestic Britain was convinced of its capacities to engage feelings of shock and awe from Indian cultures as Johnson's *Rasselas* indicates, just as often the reverse was true.

The letters written by the English in India throughout the eighteenth century convey a sense of the magnitude of an ancient and mysterious knowledge. Company servants used analogical models of reason that engaged the paradigm of alchemy. While these writers continued to record their marvels, invoking the sublimation of mundane substance into the abstraction of scientific discourse, another kind of apocalyptic expansion, however imaginative, was on the cusp of realization.

When the Treaty of Utrecht effectively ended the War of Spanish Succession in 1713, Britons found themselves faced with the possibility of accruing other new territories. The negotiation of expanded trade monopolies between Spain and the Grand Alliance (of which Britain was a principal organizer) resulted in the prospect of valuable British holdings in the New World, however abstract. In reality the British only received a share of the *asiento*, and even if this later proved profitable, at the time these prospects hardly aligned with Pope's vision of the Pax Britannica. The concessions Spain made to Britain also provided potentially vast new spaces that needed to be accounted for within British taxonomy. The *asiento*, similar to a patent in early modern England, was a trade relationship that established a monopoly for a group of traders over the trade routes or products or both. For Britain, this monopoly had to do exclusively with the trade of African slaves from the Caribbean to Spanish mining colonies. This monopoly, given to Britain by the Dutch, was a compensation for the victory of the French candidate Philip V of Spain. Hence, the Pax Britannica, which hailed a new world order according to Pope, while eventually proving quite lucrative, was certainly a treaty inundated with the problematic politics of the slave trade from its inception.

Alexander Pope's *Windsor-Forest*, started in 1704 as a pastoral and augmented in 1713 as an explicitly political lyric, celebrated the Pax Britannica promised by the signing of the Treaty of Utrecht. Many accounts of this poem focus on reading Pope's rhetoric of commodity fetishism to supply his pastoral representation of the new age of English nationalism and British imperial power. Laura Brown has argued that the poem describes the effects of mercantile expansion even as it dismembers the effects of imperial violence.[56] These readings do not question the historical validity of such an assumption. *Windsor-Forest*'s poetic rhetoric may also be determined by the same kind of British uneasiness about Asian science and technology characterizing many of the letters to the East India Company's office in London, or to the Royal Society. Such anxiety, I argue, is structured by the paradigm of alchemy in order to understand the strangeness of a new world that, for Pope, is simultaneously about addressing domestic exoticism.

The commodities featured in the poem, take shape primarily in the colors of precious goods (in Brown's words, "nature dressed to advantage with the colors of imperialism").[57] These poetic colors are wrought from textiles or dyes, from the oaks that supply navies and merchants with ships, from the blood that supplies, however metonymically, the bodies engaged in military conflict (most notably, although not exclusively, the armies fighting the War of Spanish Succession). Such blood eventually is the substance that dyes a previously transparent "Iber's Sands, or Ister's foaming Flood" an opaque red. The material origins of imperial colors—textiles, dyes, oaks, and blood—are equally opaque, alluding, perhaps to the murky history of imperial violence Pope uncovers in his celebration of a transparent and abstract peace.

Pope's description of a pastoral that succeeds the "dreary Desert and gloomy Waste" of "Ages past," despite being saturated in sylvan violence, recalls most insistently the progression from base material (desert waste) to bucolic plenty, even if it is rendered "thoughtless" because this period pre-dates Enlightenment reasoning. The "Tyrian dye" coloring pheasants and fish gives way to "The silver Eel, in shining Volumes roll'd, / The yellow Carp, in Scales bedrop'd with Gold, / Swift Trouts, diversify'd with Crimson Stains" to "Pykes . . . of the wat'ry Plains." The progression of the trope follows an alchemical telos. The mundane substance constituting "Tyrian" dyes, made from mollusks, transforms this landscape into something rich and opaque: the silver and gold that represents, albeit via the "crimson stains" that necessarily attend colonial conquest, wealth pouring in from new and old worlds to enrich the once-barren English plain. Although the endpoint might seem to rest in silver and gold—the solidity of bullion that ensures British trade—these tropes don't conclude Pope's description of the "genial Spring" or the "patient Fisher." Rather, those plains, so lately drenched in the opaque colors of imperial possession, give way to something much more transparent: the "wat'ry plain" (ll. 135–147).

Likewise, later in the poem, the "Man whom this bright Court approves" is one who "gathers Health from Herbs the Forest yields, / And of their fragrant Physick spoils the Fields." Such a man, well-versed in medicinal art (part of alchemical science), is able to determine from a natural base the spoils that he may "with Chymic art exalt the Min'ral Powers, / And draw Aromatick Souls of Flowers" (ll. 235–244). Demonstrating a singular talent to "exalt" or distill phenomenological "powers," he

Now marks the Course of rolling Orbs on high:
O'er figur'd Worlds now travels with his Eye. (ll. 245–246)

The lines from *Windsor-Forest* I have used here follow the process of distillation, but they don't conclude with opaque composites ("Tyrian dye") or distillation ("min'ral powers," "aromatick souls of flow'rs"). Instead, Pope sublimates these commodities to something more abstract and transparent. The charts of "rolling Orbs" or "figur'd Worlds" makes physical travel irrelevant; the approved man, well-versed in empirical powers of reason, may now travel, "with his Eye," toward a new cartographic comprehension of the world. Similarly, the "sons" that "dye with British Blood/Red Iber's Sands" sublimate to a "fair Peace! from Shore to Shore" that is negotiated by the "liquid" (transparent) figure of Father Thames, the new reigning authority.

But this comprehension depends on his knowledge of a premodern, pre-Enlightenment past:

Of ancient Writ unlocks the learned Store,
Consults the Dead, and lives past Ages o'er (ll. 247–248)

Drawing on the works of antiquity, this man finds his learned "store" in an Augustan and Diocletian past. He has been approved by the same court that has negotiated the Pax Britannica, which in turn has ushered in an age of "home-felt Quiet," "Successive Study, Exercise and Ease," uses an older technological paradigm in order to charter new spaces. The "rolling Orbs" or "figur'd Worlds" only make sense when a "learned Store" is unlocked, when "past Ages" are lived again. When Enlightenment analytical modes of reason fail, consulting the "Dead" from "past Ages" yields other ways of accounting for domestic mysteries. Pope uses the mythological history of England to represent Britain's changing global relations. Bringing it home, so to speak, replicates the ways in which travelers to foreign lands like India didn't simply fold into their extant language an account of foreign oddities but integrated these differences deep into the heart of their epistemology.[58]

Pope's return to other histories, however, signals more than just a need to restore or retain notions of a *prisca scientia* to pristine scientific knowledge. In his rhetorical move to account for a muddied past, to find a language to assert a new episteme or, perhaps, to find a new logical authority that is both "imperialistic" and deeply uncertain about its own principles, he invokes the problems with forms of knowledge with which Newton, among others, was struggling. For example, Newton questions the capacity for mathematics to function as a totalizing methodology even if Newton increasingly argues for the use of mathematics as a significant language with which to explore, as Robert Markley has pointed out, the complex connections between history and theology. Markley argues that Newton's corpus is marked by a series of ostensibly contradictory moves that ultimately disrupt a progressive notion of history. Newton's "redefining of order takes a variety of forms: attacks on systematizers, from Leibniz to Athanasius; an obsession with origins, which he equates with a notion of pristine, uncorrupted meaning; and efforts to defer the kind of authoritative claims for his work that were often made for him by his eighteenth-century followers."[59]

The dialectical structure of epistemology is made possible by alchemy. Pope's poetic project of *Windsor-Forest* embodies and articulates this structure: first written as a pastoral exercise and later revised as a celebration of the Pax Britannica, the political work being enacted in this rhetoric, including the invocation of Virgil that attests to Granville's commission, reflects the problems with representing a coherent national history produced from disjointed memory. Whether or not Pope self-consciously fashioned the Pax Britannica as a mask—both of past domestic violence and a future of further brutality as Britain prepared to fully immerse itself in the slave trade—it remains that the opaque, dark, and bloody figures inhabiting the poem are transformed by Father Thames's mighty "flood": the transparent and abstract discourse of Enlightenment Britain, sublimating notions of conquest into ones of the social mission Johnson mobilized in his Oriental tale.

Pope was attendant to the generic anxieties of using the pastoral to resignify a civic, national, and cultural understanding of London as a free port, a dutiless place to which all free nations naturally flow. His poetic task raises similar problems of a totalizing poetic discourse that has to redeploy an account of the past to represent a future whose parameters were less clearly defined. Unlike the grand imperial trajectory Victorian England might have imagined for itself (although clearly not unmarked by similar anxieties), Pope's representation of liquid assets is speculative. He may have had to refer to a past alchemical "truth" to solidify future projections, and the pastoral in *Windsor-Forest* is thus figured as an alchemical history.

Alchemy crucially supplied Pope with a poetic language with which to read new spaces and an ancient authority for new knowledge. It was also a language that asserted an epistemological and imperial authority with a good deal of ambivalence. This model may account for the conflicting, oxymoronic insistence on opacity. Such opacity takes shape most visibly as the commodities (imperial colors) littering the poem. These commodities are, in turn, sublimated through the alchemical discourse already saturating representations of another soon-to-be imperial possession, India (as the letter-writers demonstrate) into transparent, enlightened scientific, geographical, racial, and poetic truths. Thus the "fair peace" reigning from "shore to shore" addresses the language from one "shore" (India) to account for the alterity of another (the New World).

Alchemy provided a discursive structure with which Britons could imagine or shape their material historical engagement with cultural and epistemological difference. Alchemy, attendant to the purification of muddied substance, was a useful paradigm by which the English transformed a dark and impure land into pure wealth and the refinement of pure possibility. More crucially, alchemy displaced the necessity for accountability onto the notion of "providence." The messy, muddy reasons to represent trade or colonial conquest as divine right to a nation equally invested in the increasingly divergent discourses of imperial commerce and republicanism are simplified or purified. When Rasselas poses his question, Imlac can confidently reply, against all visible reason, but surely aware of another discourse that sanctioned the glaring logical holes left by the first paradigm, that an "unsearchable will" of a "Supreme Being" was reason enough. But early modern East India Company correspondence makes those holes insistently visible to those who want to recognize them.

In this chapter I have argued that the British tried intermittently to harness material and technological substances and methodologies to supply their own scientific gaps. Such accruement suggests that somehow these forms of knowledge were extant without human agency. Certainly the entrenchment of colonialism in the nineteenth century as a political force to keep open profitable mercantile exchange privileges the idea that Indians were only present in this process of abjection to supply brute labor, to become as inanimate as the materials their culture "offered" to their conquerors, and as unconnected to the epistemology they contributed to the interests of Western science as farm animals are to agricultural production.

Mortar and the Making of Madras

The best Sort of Whitening Varnish is thus made. Take one Gallon of Toddy, a Pint of Butter-Milk, and so much fine *Chinam*, or Lime, as shall be proper to colour it; add thereunto some of the *Chinam* Liquor before mentioned, wash it gently over therewith; and when it is quite dried in, do the same again. . . . The Plaistering above described, is thought in *India* vastly to exceed any Sort of *Stucco*-Work, or Plaister of *Paris*; and I have seen a Room done with this Sort of Terrass-Mortar that has fully come up to the best sort of Wainscot-Work, in Smoothness and in Beauty.

— ISAAC PYKE, *The Method of Making the Best Mortar at Madras in East India in Philosophical Transactions of the Royal Society* (1731)

[Madras] is, without exception, the prettiest place I ever saw. Madras is built entirely by the English: it is strongly fortified, and the walls and works, as well as the barracks for the army, the storehouses, and every other public building are so calculated as to be both convenient and an addition to the beauty of the place. . . . The *varendars* give a handsome appearance to the houses on the out-side, and are of great use, keeping the sun out by day, and in the evenings are cool and pleasant to sit in. But what gives the greatest elegance to the houses is a material peculiar to the place: it is a cement or plaster called *channam* made of the shells of a very large species of oysters found on this coast: these shells when burnt, pounded and mixed with water, form the strongest cement imaginable: if it is to be used as a plaster, they mix it with the whites of eggs, milk, and some other ingredients: when dry, it is as hard, and very near as beautiful as marble.

— JEMIMA KINDERSLEY, *Letters from the Island of Teneriffe, Brazil, the Cape of Good Hope, and the East Indies, Letter XIX* (June 1765)

Isaac Pyke and Jemima Kindersley's visits to Madras were made thirty-four years apart from one another; yet they were both struck by the same phenomenon. Both of them observed and then reported on the extraordinary quality of the mortar and plaster that was commonly used to construct the city's buildings. Pyke sent his notes to the Royal Society, England's most respected scientific organization, and was subsequently published in its *Philosophical Transactions*. Kindersley didn't have quite the same resources at her disposal as had Pyke, who was then the governor of St. Helena, but she was an intrepid and perceptive traveler and her letters, published in 1777, were valued as a meticulous record of East India Company government in India.[1] Kindersley includes the procedure of creating mortar as one of the many things that make Madras "without exception, the prettiest place I ever saw," and Pyke elaborates on the different forms of mortar and plaster and includes several recipes in his treatise. But Kindersley's remarks on the general appearance of the town include a detail that Pyke glosses over: the most important ingredient to the mortar and plaster, *chinam*, itself comes from the shells of oysters indigenous to this coast that are "burnt, pounded, and mixed with water."[2]

Pyke and Jemima Kindersley didn't explicitly connect mortar and plaster to the process of empire-building. Pyke was absorbed by his gubernatorial duties on St. Helena and Kindersley was en route to Calcutta to join her husband. Yet they both thought the mortar and plaster were worth mentioning in their journals. I make this connection explicit by establishing the multiple ways in which the humble but nevertheless material mortar sublimates into the abstract ideological project of building an empire.

Mortar

The first English holding on Indian soil was a place of radical experimentation for the East India Company. Purchased by Francis Day in August 1639 from the reigning king, Peda Venkata Raya, Fort St. George emerged from the small strip of land as a factory and warehouse for the trading enterprises of the Company, which had finally obtained a *firman* from the Vijayanagara sovereign. The early years of the settlement, like English settlements all over the world, were characterized by a good deal of hardship, strife, and fierce competition. Added to this general frustration was the knowledge that their presence on the lucrative Coromandel coast was tenuous at best, and it made sense to give due deference to the powerful rulers of the area.

But by the beginning of the eighteenth century, the British were self-consciously consolidating control over people that they had renamed "black"; in the case of Madras, this city had become the "first city in world history to designate separate sections of the town by color, renaming its 'Gentue Town' or 'Malabar Town' as 'Black Town,' and its 'Christian Town' as 'White Town.'"[3] The divisions between these sections of town were both architectural and ideological. Solid walls made of mud and clay surrounded the town, but they also divided it in two. These walls were not as impermeable as they appeared to the Court of Directors in London, not from their first erection to their later solidification. And before the renaming of Christian Town and Gentue (Hindu) Town to White Town and Black Town the differences between these sites were not obvious: they had to be invented, administered, and policed.[4]

Company agents spent the first hundred years or so of Madras's existence maintaining and refining this wall, building new walls and constantly pleading for money from the Court of Directors in London to secure them.[5] The walls, however, were not enough in and of themselves to separate Christians and "Gentues." One of the earlier governors of Fort St. George, Sir William Langhorn, worried extensively about them. In the 1677 he wrote the Directors in London:

> The outward wall of the House in the Fort being found to be very crazy and tottering through badness of the foundation, and many cracks more and more appearing therein, upon which surveigh by the Chief Gunner and Ingenier of the Fort, Wm. Dixon, Muttamarra the chief Carpenter, and Nallana the chief Bricklayer, has already obliged us to run up two Buttresses the last yeare to the North East, and one this yeare to the S.W. to support it; and these not suffising, it is found needfull to set up one Buttress more to the S.W. and one more to the N.W., three more to the N.E. and two or three Timbers to the S.E. where the Jettys hinder the joining of the Buttresses, to prevent if possible the suddain falling of the said wall, and for the security of the lives of the Agent and others of Councell, the Minister, and Factors and Writers

inhabiting therein, which, upon any great storm of wind and rayn (very frequent in these parts), are in very great danger, the whole house rocking in a strange manner, and built of Brick and playster without lime; the Agent and Councell being desirous to tarry if possible for leave from the Honble Company and for taking of it down and new building it, the necessity whereof is to be represented to them by these letters.[6]

Langhorn's description of the precarious House with its "tottering" structure and unsound foundation articulates quite eloquently the ideological issues his soldiers, factors, merchants, agents, writers, and ministers faced. In another plea to the East India Company Directors, Langhorn and his council emphasized the "thinn, low, slight, tottering Walls, as already advised the Honble Company, Pestered with a great Town close to them."[7] Far from the stout walls that later were to separate Black Town from White Town, this one was only a cipher of a wall—"thinn, low, slight, tottering"— hardly the kind of daunting defense the Company servants needed against the "Pestering" from the "great Town close to them." Langhorn devoted himself to the fortification of Christian Town and sought to uphold topographical and cultural barriers even while allowing Indian labor to assist with construction: the chief carpenter and bricklayer (Muttamarra and Nallana) were both from the "great Town" that was the native city of Madras. As an example of the term that historian Carl Nightingale coins as the "servant-exception rule," Indian labor could move between the two sides of the wall, and as the town and its needs grew, even live on the English side of the wall which, despite its semi-permeability, maintained the ideological divide as long as the labor was servile.

The Fort House "rock[ed] in a strange manner." It was built of brick and plaster but, like the wall, it was unstable and subject to sudden collapse during any one of the frequent storms or seasonal monsoons. Clearly, something was wrong with this British edifice. Langhorn had asked for permission to demolish the existing building and replace it with a new one, particularly noting the absence of native lime in the plaster of the original structure. Isaac Pyke, twice governor of St. Helena (1714–1719; 1732–1738), was a keen observer of local knowledge. Writing to the Royal Society in early 1753, he remarked on the making of mortar in Madras and offered several recipes he had collected from Black Town artisans for different mortars, including one for "when the Work is intended to be very strong; as for Instance, Madras's Church Steeple, that was building when I was last there."[8] All these recipes included liberal amounts of *chinam* or lime used in both mortars and plasters. Both Isaac Pyke and later Jemima Kindersley were struck by the almost mystical strength and beauty of this mortar and plaster, composed of local ingredients and used by local builders. Pyke comments on its "smoothness and beauty" while Kindersley extols the "greatest elegance" the plaster gives to the houses. Like Pyke, who saw nothing wrong with making abundant use of local techne, Langhorn identified the failure of incorporating local techne in establishing a settlement. Constructing the Fort House with bricks and plaster "without lime"—without the technological ingenuity wrought by Indian bricklayers and carpenters—only resulted in a "crazy" structure, whose shakiness hardly represented a permanent English presence.

Trickier for Langhorn was managing other forms of cultural exchange that occurred despite the wall's presence or, perhaps, because of it. Embedded in the correspondence between the Court of Directors in London and Langhorn at Fort St. George is a serious disjunction. In a response to complaints made by the Directors, Langhorn attempted to regularize British behavior. He drew up a prescription of orders in 1678. Among those orders were the following:

7. It is likewise ordered that both Officers and Souldiers in the Fort shall on every Sabbath day, and on every day when they exercise, weare English apparel; in respect the garb is most becoming as Souldiers, and correspondent to their profession; in penalty of forfiting one months allowance on the Officers part, and half a months allowance on the Private Souldiers part.

8. Whosoever he be that shall attempt to get over the walls of the Fort upon any pretence whatsoever, shall for so heinous and grievous an offence be kept in Irons till the shipps arrival; and then, his wages being suspended, be sent home for England, there to receive condigne punishment. . . .

10. That when the Governour, &tc., shall go on board, or abroade on horse-backe or in pallenkeen, it is thought fit, in respect of the small number of people, that not a man shall stir out of the Fort until the Governour returns home; upon penalty of half a riall of eight for the Merchant and Officer, and a daye's sentinel in armes to the Private souldier.[9]

It would have been unthinkable for nineteenth-century British soldiers and officers stationed in India to be seen in any garb other than English, but by then the British had co-opted the textile industries of India, which produced a much finer cloth for uniforms.[10] Officers and soldiers alike, dressed in the coarse woolen broadcloths of seventeenth-century English manufacture, would no doubt have been tempted by the high quality of the muslins, calicoes, cottons, and other finely woven textiles that suited the climate so much more appropriately. Likewise Indian fashions, admirably designed to combat the climate, were equally tempting to the badly dressed members of the East India Company. They were obviously spending their hard-earned (Spanish) cash somewhere: nipping over the wall to Black Town to purchase clothes and other goods, to drink at any of the numerous punch houses, or to engage in sexual interactions with Indian men and women.[11] As J. Talboys Wheeler notes in his 1878 *History of English Settlements in India*:

> The neighbourhood of Black Town was not conducive to the morals of the Fort. The younger men would climb over the walls at night time, and indulge in a round of dissipation. There were houses of entertainment known as punch houses. . . . They took this name from the Indian drink concocted by the convivial Factors at Surat.[12]

Semi-permeability may have been built into the wall, so to speak; nevertheless, Langhorn's decree indicated that there was trouble in the British community of White Town. In fact, many of the orders were designed to keep denizens *inside* the walls of the Fort that marked the boundary between Christian Town and Gentue Town. Ironically, nothing short of the threat of binding miscreants in chains until the next ship sailed back to England seemed to work. Langhorn went to a great deal of trouble to remind his charges of what it meant to be a English soldier—the "English apparel" being the "most becoming" to the profession.

Pyke referred to St. Mary's Church, which was the first Anglican Church built in India. This structure, made with the mortar that Pyke observed, is, as H.D. Love has described, "raised in the most solid and substantial manner. . . . There has been no settlement of the foundations, and the fabric is as sound to-day as when it was built. . . . There is no older masonry structure in Fort St. George than St. Mary's."[13] The walls of the Fort were prone to collapse if they were built without *chinam*. The walls of St. Mary's, while no less ideologically fraught, nevertheless survived not just the wear and tear of time but several assaults in the late eighteenth century because they were built on the solid foundation of local "Gentue" knowledge. Langhorn's orders to confine his garrison to English apparel no matter how badly suited to the weather, and to the walls of the Fort no matter how rickety, seem as "crazy" as the original Fort House: full of cracks and flaws, and liable to fall to pieces if care were not taken to integrate local knowledge.

The early records of Fort St. George thus represent a settlement whose Englishness was difficult to secure. As the first English holding in India, the fort and its inhabitants (with the exception of the governor) seem more inclined to integrate into the "Great Town" than to cordon off a specifically Anglican space for themselves. Exchanging their woolen uniforms for the cooler, more comfortable cottons and muslins for which Madras was famous, English soldiers and officers alike had to be forced by direct order to display their national identity at least once a week. These orders were not the result of local governors but, rather, were directives coming from London. The disjunction between what the home office imagined life in India to be like and the quotidian realities Company agents faced, already articulated in the very early correspondence in the beginning of the seventeenth century, had not ceased even at the turn of the eighteenth century and the renaming of Gentue Town and Christian Town. Fort St. George could not remain erect without the incorporation of Indian techne; the bricks and plaster of early English manufacture—made without the Madrasi technique and the crucial ingredient, *chinam*—resulted in a frail structure that barely lasted thirty years, a sad testament to English technological ingenuity. The beauty of White Town that Mrs. Kindersley noted in her letters was largely the result of Indian technologies: the *varendars* (verandas) she described on the houses of White Town were distinctive to Indian architecture, designed to stave off the afternoon heat. Likewise the mortar and plasters, which were also characteristic to Madrasi building and were famous for their startling whiteness, were now used in the building of English houses. Ironically, then, the techne from Black Town sustained the ideological dominance of White Town, even to the detail of its literal and figurative color. Of course, the houses of Black Town were just as white as those in White Town, which meant that the "blackness" of Black Town—the economic, political, and eventually racial subordination of Indians—had to be continually reinforced. Nightingale describes some of the ways in which this was done: "Proclamations from the Agent or Governor were traditionally issued by cannon shots and a loud procession that passed from the White Town through the massive Choultry Gate into the Black Town. All of those ordinances were later posted on the huge wooden doors of the same Choultry Gate. White Town was a grand stage for the theatrics of imperial authority."[14]

Unlike other early English settlements in the United States and in the Caribbean that tried so hard to replicate English towns, Company members in Madras were far more prone to assimilate into Indian culture.[15] This tendency created a problem for Company hegemony; governors and soldiers had an annoying penchant for personal gain rather than strict adherence to the Company's interests. They formed their own trade relations with prominent and powerful Indian merchants in order to share in the profits of trade: Madras was the center not only of valuable textiles but also of the diamond, spice, and slave trades. The pageantry Nightingale describes seemed to be far more necessary for the inhabitants of White Town than those of Black Town, reminding them what it meant to be English. In fact, the divide between White Town and Black Town was legal and ideological but *not* material; laws that declared this difference were necessary precisely because there was no architectural difference between the two parts of town. The English were, in effect, living in an Indian city no matter what they named it.

Two years before his tenure as governor was terminated, Langhorn had written to the Directors of his hopes and plans for a Garden House being erected, for ceremonial purposes. They agreed and ordered in 1676

> that the said House or Choultry be built accordingly, and that it may be
> handsomely built and of proportionable size, large enough to receive
> Phirmaunds from the King and persons of quality, the design for which it
> was intended and desired; and yet so contrived to be no great Charge to the
> Honble Company.[16]

The distance between the East India Company's Court of Directors in London and its agents, merchants, governors, factors, and writers in India was great. At least a six months' voyage by ship, correspondence between the home office and members posted in the East was often completely outdated by the time it had reached its recipients. But even more entrenched was the cultural distance between the Directors and Company members in India. First, there was a constant struggle over money: what things cost and what people imagined they cost were two separate ideas. The "handsomely built" Garden House of "proportionable size" was somehow to be erected at "no great Charge to the Honble Company," even in the face of Langhorn's steady pleas for more funds merely to reinforce the walls of the Fort let alone think about its ornamental possibilities. Even more striking, however, is the fact that the Directors imagined the Garden House as a place "to receive Phirmaunds from the King and persons of quality." The only binding *firman* the East India Company had received on the Coromandel Coast was the one Francis Day managed to obtain in 1639 that enabled the English to establish themselves in Madras. That the new Garden House would, in the eyes of the Directors, be a suitable place to receive scores of "phirmaunds" suggests that profitable English trade with the sovereigns of the Indian Ocean was still largely an English fantasy. The Garden House was not built during Langhorn's tenure. In 1678 he was accused of private trading by the Directors and was relieved of his command. He returned to England, where he purchased an estate, Charlton House, in Kent, and went on to endow schoolhouses and almshouses and in general to perform good works, thus discrediting the Company's accusations of personal greed.

The charge of private trading leveled against Langhorn was not an unusual one. Private trading was the one certain way Company members had to secure personal fortunes and interlopers, as they were also called, were well represented in Fort St. George's gubernatorial roster.[17] What is more interesting is to consider the effects of private trading. Passing through the walls the English built was a form of intellectual and cultural exchange—an acknowledgement, tacit or not, that this small settlement had a good deal to learn from the vast Vijayanagara, Maharatta, and Mughal empires. Textiles, diamonds, and spices were all very well and good but equally valuable to the English living in India were technologies that enabled them to acclimate to their environments. It was only through private trading that the English acquired such techne as *chinam*.

Mortar, then, is an overdetermined substance, a composition of material things that sublimates to something with unimaginable value. Isaac Pyke observed the various methods of mortar making in Black Town, and sent his comments to the Royal Society, being careful to note, "whereas sundry ingredients here mentioned are not to be had in *England*, it may not be amiss to substitute something more plentiful here, which I imagine to be of the same nature." He then offered "Oaken-Bark" to substitute "astringent Barks," the "Bark and Branches of the *Sloe*-Tree" for "*Aloes*," "*Molasses*" for "*Jaggery*," and "Liquor from the *Birch*-Tree" for "*Toddy*."[18] Pyke's paper is a painstaking study of Indian techne that he saw fit to share with the major scientific organization in England, which suggests that this was not simply a whimsical exercise. Like many travelers to India, Pyke's helpful analogies indicate a willingness to integrate foreign knowledge into more familiar materials for his English readers. Thus, forms of Indian knowledge are transported to England and lodged in the annals of British scientific discovery. But even more materially, Madrasi mortar became the substance that supported the British structure of power: the architecture of the Fort that communicated, according to Nightingale, such "commanding superiority" was held—even stuck—together with Indian techne, even though the ideological distance between the British and Indians became greater than ever.

If Madrasi mortar was the material substance that bonded this small English settlement, it was also the stuff that allowed the English to break away from their dependence on Black Town and imagine their space as substantively different. That is, once this mortar secured the buildings within the Fort, it was easier to picture an existence separate from Black Town. English incorporation of Indian techne into their own practice enabled observers like Jemima Kindersley to remark that Madras was "built entirely by the English." Mortar's material capacities sublimated into an ideology of bifurcation that was crucial to enabling this small, vexed, and unimportant purchase to emerge as a dominant imperial space.[19] White Town and Black Town consequently became places impossibly distant from one another. Subsequent representations of the British Raj, particularly in the nineteenth and twentieth centuries, honor this divide and concentrate on the ways in which British techne reformed India, for better or for worse, and relegated it until very recently to the confines of the underdeveloped Third World.

Interrogating this divide makes it possible to apprehend its emergence, and grasping its imaginary validity (and not its material one) exposes the fallacies of Western exceptionalism. The genealogies of mortar, plaster, and the spaces they built also identify a different kind of history, one that situates global mercantile capitalism as a critical link between diverse cultures, traditions, products, and techne.

The Making of Madras

William Langhorn was only one of a series of governors of Madras who profited from private trade and was deposed by the Company. Returning to England, he enjoyed his ill-acquired gains in part by laundering them through philanthropic acts. His money, however, remained in England and accrued to the general economic health of that country. In the eyes of the East India Company's Court of Directors, Langhorn atoned for his dubious behavior as governor and helped to sustain the ideology of bifurcation. Elihu Yale, one of Langhorn's successors, did not. He is a figure whose tenure as governor of Madras seems impossibly distant from his later fame as the founder of a major university. Unlike Langhorn who kept close to home, Yale engaged in private trade that had far more profound consequences that complicate the divisions and connections between the Atlantic and Indian oceans.

Environmental criticism and environmental history may help excavate these divisions and connections from the profound collusion that sustains the fantasy of a single imperial history. In his discussion of comparative and transnational histories, Richard White contends that history as a discipline is the "child of the nation-state," explaining that since "the nation-state took the stage in the early modern era—or rather once the nation-state became the stage, historians have almost automatically used it as the preferred scale for their work."[20] Company directors in London were anxious to establish a specifically English place in the lucrative traffic of the Indian Ocean and their correspondence with the factors and governors in Madras takes shape as the story of British intrepidity. But correspondence goes both ways, and reading the letters of the factors and governors which included representations of mortar and other local techne, changes the grounds of the story, turning into one that is far more global than British.[21]

It follows then that reading the incidents in Madras as transnational ones complicates the boundaries of the nation-state and invites us to use global history as a more useful template. Because histories are largely wrought as matters of scale, simply replacing one with another doesn't necessarily lead to a more accurate representation. The "history of the nation has involved the subordination of any entity operating on a smaller scale," which is to say that regional and local histories are frequently either expunged in the service of national histories or treated as their metonyms. But, as White continues, even if the "global has emerged as a real and important space," to "allow it to erase other social spaces and scales, as the national did before it, would be a serious mistake."[22] Rather, White calls for an awareness of interlocking scales that are themselves socially and historically produced. In that spirit, then, my representation of Elihu Yale connects histories of the objects constituting his gift to the college—the "books, Indian textiles, and other goods"[23]—with the various local spaces Yale occupied, regions on the verge of being established as nation-states.

I begin my representation of the divisions and connections Yale embodies, as disconnected as it may seem, with one of the "other goods" of his bequest: nutmeg. Yale's wealth came from his private trade with powerful Indian merchants, and his warehouses in England where he eventually returned were filled with diamonds and other precious "wearing jewels," textiles, leather goods, oriental screens, and spices.[24] I am especially interested in nutmeg because of the qualities peculiar to the plant that made it "the most coveted luxury in seventeenth-century Europe, a spice held to have such powerful medicinal properties that men would risk their lives to acquire it."[25] Like diamonds, nutmegs were small and highly portable. They were organic and unprocessed and, unlike diamonds and textiles, could be ground and mixed with other spices to concoct other valuable substances. They were embodiments of surplus value *in potentia*, and Yale made sure to include many "jars" of this spice in the cargo of the East Indiaman, the *Martha*, on which he finally sailed back to England.

As an object, the fruit of the nutmeg tree, *Myristica fragrans*—a small, unassuming, round brown nut—has had, at various times, substantial histories attached to it that increase or decrease its value in Europe and the United States. This object has been variously prized for its use as a spice, a preservative, and a medicine, and has also been central to the subject of economic competition and colonial warfare. In yet another capacity, this object has been crucially forgotten: another jar on the spice shelf gathering dust, which has transformed into a subject of (national) complacency. As a coveted spice, it radically shaped the identity of Enlightenment Europe and the New England colonies that eventually became the United States.[26] Nutmeg played a critical part in institutionalizing Linnaean taxonomy, European epistemology, and, finally, American academic hegemony.

Nutmeg's move from high visibility as a valuable commodity with a plethora of capacities to relative invisibility as simple spice is not a result of the inevitable march of progressive national history. Such a history imagines a central narrative with a central theme that inevitably results in eradicating ostensibly anomalous, marginal, or peripheral material. That is, if one removes nutmeg from the center of the narrative of Europe's entrance into global trade, it is relegated to the back of the spice shelf. Assimilated by the intrepid advance of European commerce, which culminates in the grand narrative of Enlightenment pre-eminence, nutmeg is forgotten, only brought out when the occasion demands its use, reminding us of yet another commodity enriching the national larder. If nutmeg is itself the central narrative, then multiple and often competing histories emerge that render accounts of Western exceptionalism like *Rasselas* largely the result of its self-image.[27]

It is the latter paradigm I want to use to reconceive national representation as transnational phenomena. Nutmeg happens to share a history with Elihu Yale whose personal life, colorful as it was, has since yielded to its own institutionalization. Nutmeg has a long and complicated relationship with several European trading companies, most notably the Portuguese and the Dutch, but also the English and the Americans; and its specific connection to Yale, although apparently tangential, turns out to be crucial to uncovering the divisions and connections between the Atlantic and Indian oceans.

Growing only under the extraordinarily specific conditions that the microclimate of the Banda Islands provides, nutmeg entered British cultural consciousness through the marketplaces of Constantinople and Venice. Nutmeg had been prized as a preservative as well as a spice, but its value increased exponentially when Elizabethan physicians decided—perhaps because of its rarity—that nutmegs cured the plague. Thus its medicinal properties acquired mythical and mystical status that goaded English merchants to brave the hazards of Portuguese and Dutch competition and their own dim understanding of the Indian Ocean in order to find the origins of this spice.[28]

The entrance of nutmeg into European sensibility is compellingly told in Giles Milton's popular history, *Nathaniel's Nutmeg*, and more academically engaged by environmental historians Richard Grove and E. C. Spary.[29] Milton's riveting tale traces the origins of the British entry into the nutmeg trade in pursuit of its profit; he is thus invested in participating in the grand narrative of British and European commercial ascendancy. Grove and Spary are concerned with the history of another aspect of this spice: the politics of scientific conservation and botanical knowledge that created a space for nutmeg in European taxonomy.[30] Their studies of colonial botany contribute less to the discourse of Linnaean taxonomic pre-eminence than to the complex political maneuvers British and other European nations were forced to engage in order to have any purchase whatsoever in the lucrative spice trade that had hitherto been monopolized by nations of the Indian Ocean.

Grove outlines the ways in which the French "examined the possibilities of growing spices in their tropical territories, especially nutmeg and clove plants." The French were particularly interested in annexing Mauritius (which they ruled from 1722 to 1790) in response to their competition with Britain to "build up spheres of influence in India and the East Indies." Like the British, the French had a less than secure audience with the powerful rulers of the Mughal Empire and the lucrative marketplaces they controlled. With the Dutch hegemony of the European spice trade, it made a good deal of sense for the French to set up alternate potential plantations in order to compete effectively in European markets. Under French rule (after being abandoned by the Dutch), Mauritius "became the location for the flowering of a complex and unprecedented environmental policy," mostly to displace the Dutch monopoly of the spice trade. According to Grove, the French had considered growing nutmeg and cloves in their tropical holdings, and the "perceived potential of Mauritius as a site for spice plantation . . . is confirmed by a letter written in 1716 by Dulivier, the governor of Pondicherry."

Mauritius "occupied a strategic position in French thinking and soon invited the prospect of direct intervention by the French Crown." Mauritius thus transformed from the material to the abstract: from a geographical site, a botanical experiment to further French purchase in the spice trade to a "strategic position" in European consciousness.[31]

E.C. Spary's approach to the nutmeg phenomenon also addresses the cultivation and conservation of the environment for the production of nutmeg. Her method focuses on the transmission of nutmeg from its East Indian material origins to the intellectual property of French physiocrats: how nutmeg simultaneously operates as spice and knowledge claim. Spary focuses on stabilizing the identity of nutmeg through the mastery of Linnaean classification. "For a plant such as nutmeg to be the subject of scientific experimentation and agricultural exploitation," she writes, "it had to be assimilated into existing classificatory and descriptive schemes, yet clearly distinguished from similar species." This particular scientific process was necessary even after nutmegs had entered European discourse and had been established by European botanists because only "these social and medical practitioners could stand as credible guarantors of authenticity of a particular species." Here Spary is using Steven Shapin's arguments for the credibility of a knowledge-claim to address the care with which Pierre Poivre—the physiocrat who was responsible for introducing forest and climate protection measures to French colonial holdings in the East Indies—took to authenticate "true" nutmeg.[32] With the Dutch monopoly on the nutmeg trade firmly entrenched (and the measures they took to ensure high prices, going as far as destroying the orchards that produced them on other islands), it behooved the French to try cultivating the plant elsewhere—a measure that eventually proved successful. The problem was how did one know that one had a "true" nutmeg and not some interloping hybrid? Spary argues:

> In part, identity was stabilized through the mastery of classification. Natural history supplied preexisting frameworks of description and differentiation: by slotting a given natural body into taxonomic categories, one could fashion its scientific identity. . . . It is a testament to the difficulties we *should* ascribe to the identification process that criteria such as odor and appearance were not enough to settle the issue in the eighteenth century, even more familiar with nutmeg (then a widely used seasoning and medicament) than we are.[33]

Scientific discourse—the "identification process"—trumps simple sensory criteria like smell and form and authenticates nutmeg in all its unique epistemological value, firmly secured in the archives of institutional knowledge. The desire for nutmegs—as spices, as medicines, as immensely valuable commodities—was transformed into a knowledge of them: their primary value transformed from the material to the abstract. Driving French agronomic projects was not the production of nutmeg but the desire to produce "true" nutmegs—ones that had been scientifically authenticated. For example, the celebrated pharmacist, Jean-Baptiste-Christophe Fusée, refused to acknowledge nutmegs harvested from the plantations in Mauritius, stating "I neither could nor would acknowledge that tree and those berries to be the true nutmeg of commerce."[34] Botanists thus "simultaneously engaged in the micropolitics of botanical identity and the macropolitics of eighteenth-century colonial and social life."[35] In so doing, they supplemented the material presence of the foreign object initially resisting classification (think of the Elizabethan doctors who were so certain that pomanders made of these strange nuts would cure the plague) with encyclopedic knowledge.[36]

By the end of the seventeenth century, nutmeg moved from being a material good that operated along networks of trade, consumption, and economics to an abstraction—part of scientific discourse and botanical knowledge. Still prized for its substantial value as a commodity—a staple of everyday cuisine in one culture could, when dried and packed and shipped across oceans to another, accrue a mark-up of 60,000 percent—nutmeg was transubstantiated by Enlightenment science from a spice to a knowledge claim.[37] This claim was dependent on Enlightenment models of empirical experimentation that defined nutmeg as a natural object only once the social setting in which its meaning would be granted had been controlled. Nutmeg's place within European taxonomy was inextricably linked to its value as a desirable commodity.

Thus far I've pointed to several stories about how nutmeg functions as an event as well as a thing. But what of its relation to Elihu Yale? My thesis, as reductive as it may sound, is this: Yale is a nutmeg. Let me explain.[38]

Elihu Yale is probably most famed, at least in the United States, for funding a college that became a major university in the state of Connecticut. Yale also was the Governor of another kind of "state" in India: the

English holding of Fort St. George or Madras.[39] Such a fact is less well known in the United States although in India Yale's position in Madras (in addition to his connection with the university) is something every school-aged child knows. As the founder of a university, Yale is clearly American; as a factor of the British East India Company, his place becomes murkier, less lucidly defined by allegiances to either England or its colony in America. That is, if history is the child of the nation-state, it becomes more difficult to account for fractured allegiances.[40] My interest in Yale as a transnational phenomenon comes from these apparently contradictory positions. I'm not suggesting that one couldn't be simultaneously the founder of an American university and a governor of an English settlement; rather, I argue that these positions seem impossibly distant from one another. In other words, the event of Yale University seems utterly detached from the event of Elihu Yale no matter what discourse the university invokes about his history or his biography.[41] I want to identify the intersections between these disparate positions in order to think about other connections that are equally incongruent.

Yale was born in April 1649 in Boston but emigrated with his family to England when he was quite young. The title of Hiram Bingham's 1939 biography attempts to put together the various subject positions Yale occupied: *Elihu Yale: The American Nabob of Queen Square*. Such a title appears to attend to the discontinuities of Yale's life, but still serves to represent national identity. The blurb attached to the inside cover is equally telling: "The *Weekly Gazeteer* announcing the death of Elihu Yale did not mention that the deceased had begun his career as a penniless American in London. Today Elihu Yale's bequest has grown beyond his wildest expectations, but his own colorful career has been well nigh forgotten." This copy announces Yale as unproblematically "American" even though the New England colonies where he was born and to which he never returned were a far cry from the American nation that subsequently emerged. His tenure as an Indian "Nabob" in Madras and master (according to Bingham) of "Queen Square" in London— places where he did spend a good deal of his adult life—seem ancillary to his reputation as a great American life. Moreover, the markers of his "colorful" life seemed to have slipped off, sublimated, or abstracted into the gravitas of the university, despite the promises the text makes. Both assumptions—Yale's American identity and blameless life—only make sense if we think of biography—life writing—as a monolithic representation serving a single nation's interests.[42]

Yale set off from London for Madras in December 1671 as a young factor for the British East India Company. Most of his duties in Madras had to do with receiving and invoicing textiles, but he was confronted by the difficulty of nomenclature of the incredible variety of cloths whose names—indigenous ones—varied with texture, place, and origin, with breadth, weave, weight, fineness, stripes, or prints. He successfully overcame these difficulties, demonstrating his ability for this kind of work. More important, he was successful at obtaining a cowle (if not a *firman*) from the Harji Raja, and writing to Governor Gyfford, Yale claimed he had "preserved the Honble Companys honour and their money too . . . which makes me bold to mind you of your promised honour of meeting us at the honble Companys new Garden, where we hope to kiss your hands by five of the clock on Wednesday evening."[43] Yale managed to act out the Company's desires for the use of the Garden House, bringing back a lease for trade at Porto Novo and Cuddalore and sealing the deal in the new Garden. The Company rewarded the good behavior of its servants, and Yale was appointed governor in 1687, fifteen years after he had first come to India.

He was, however, deposed five years later, primarily for the crime of exceeding his juridical powers under the East India Company charter, but also for other transgressions against church, monarch, and charter. Mindful of the problem with interlopers, the Court of Directors obtained a new charter from Charles II and in 1683 they wrote the governor of Fort St. George, apprising him of the new terms of the charter:

> Herewith Wee send you an authentiq[ue] Copy of our new Charter granted by his Majesty for suppressing Interlopers (under the great seal), of which more by our next ships. In the mean time We appoint our Agent and Governor at the Fort to be our judge Advocate at that place, and put all the powers therein in Execution.[44]

This charter was clearly in execution when Yale assumed office; yet he flaunted its restrictions, making lucrative deals with the then interloper, Thomas Pitt (who later became a governor himself), and engaging under

Pitt's offices in a lucrative trade in diamonds and spices: cloves, cinnamon, and nutmeg.[45]

He also almost certainly profited from the burgeoning slave trade in Madras. In 1683, the Directors reported "The trade in slaves growing great from this Port, by reason of the plenty of poor, by the sore famine, and their cheapness,—it is ordered for the future that each slave sent off this shore pay one pagoda to the Right Honourable Company."[46] In September 1687, The Company again issued an order to

> buy forty young sound slaves for the Right Honourable Company, and dispose them to the several Mussoola Boats, two or three in each, in charge of the Chief man of the Boat, to be fed and taught by them; and to encourage their care therein, it is ordered a short red broad cloth coat be given to each Chief man; and that the Right Honourable Company's mark be embroidered with silk on their backs . . . whereby we shall have them at better command.[47]

Ironically, one of the accusations leveled at Yale during his last years in office was his leniency toward native criminals. "We are obliged to declare you, the Honble Elihu Yale, Esq., President and Governour of this place, answerable for all damages that have or may hereafter accrew to any Person by your unwarrantable forceing men from the hands of Justice," wrote an indignant group of aldermen, including the Scots lawyer William Fraser, who was one of the first people to accuse Yale of crimes against British jurisprudence. Yale responded that they unjustly accused him with "intrenching upon your privileges by releaceing Wassalinga, a poor miserable wretch that by the cruelty of Mr. Bridger, hath been close confin'd for debt near 12 months till he almost perisht with hunger and Sickness, without either care or relief from his merciless Creditor."[48] But Yale's interest in the lucrative trade in slaves came after he had been deposed as governor and during the seven-year tenure he spent in Madras amassing an enormous fortune. One of the pet grievances the East India Company had against Yale was the very fact that he returned to England having made a fabulous personal profit, none of which accrued to them.

While governor, he was accused of overreaching the administrative powers vested in him by the Company. Nathaniel Higginson (who later succeeded Yale as governor) together with the lawyer William Fraser, were appalled by his executive decision making, which included the unilateral resolution to use *chinam* and other local techne to fortify some of the ramparts of Black Town and White Town that were being eaten away by the sea. This in and of itself was not necessarily problematic for the Directors of the Company, but the fact that Yale resisted supplying the Company with its share of the local taxes and revenues gleaned from the call for "all coolies, carpenters, smiths, peons, and all other workmen, and all that sufficient materials be provided, that they may work day and night to endeavor to put a stop to [the sea's] fury" was very much of a problem for the Home Office, especially considering that, in their estimation, "the inhabitants . . . do live easier under our Government than under any Government in Asia, or indeed under any Government in the known part of the world."[49] Yale's interest in Madras as an *Indian* city was reflected in his role as an arbitrator of justice, his understanding of the value of Indian techne, and his refusal to sustain the Company's coffers with proceeds from his own juridical decisions. He had effectively ignored the wall.

The ease with which "inhabitants" lived under Company government referred to the dwellers of Black Town, and this claim was intended to reflect on the beneficence of their rule. Yale seems to have adjusted that claim to pattern his own erotic life. His wife Catherine, fearing for the health of her remaining children, returned to England. No sooner had she set sail on the *Rochester*, carrying with her a substantial fortune in diamonds garnered from Yale's negotiations with the most successful diamond merchant in Fort St. George, Jacques de Paivia, than Yale took up with Paivia's widow, Hieronima. Yale's brother Thomas returned from China to India to find his brother living with her and the children he had with her in the garden house. Matrimonial and religious transgressions scandalized the British Christian community of White Town who heard "a story told all about Towne of a Mrs. Nicks [who] then lived with Mr. Yale at his Garden house where she and Mrs. Paivia, a Jew, with their children have and doe frequent to the scandal of Christianity among heathens."[50]

The famous Company Garden House, envisioned by the Court of Directors as a dignified and elegant space in which governors would graciously receive *firmans* granted by Indian sovereigns, eager to prove their solidarity with English power, now housed Yale's adulterous dalliance and illegitimate

issue. The walls had come down. Far from guarding the borders and ramparts of this self-consciously structured Christian Town, Yale himself
breached their fortification by consorting with Jews and providing a putatively reprehensible model to the "heathens" of Black Town. It seemed
that Company invectives against interlopers and private trading were not
only about the loss of money; as with Langhorn, Yale's private trading
exhibited powerful appetites among East India Company governors that
had to be controlled by frequently repeated orders from the Court of
Directors.[51]

Yale's blatant disregard for the East India Company's directives, demonstrated by the openness of his association with the infamous interloper,
Thomas Pitt; the fact that he remained in Madras after he had been
disgracefully deposed; the ways in which he flaunted his sexual liaison with
a woman who were not only married (although shortly afterward widowed)
but, to the distress of this consciously Christian community, Jewish—none
of this would have been so problematic had the Company not resolved to
make Madras its headquarters in Eastern India. In 1658 the Company had
declared all of its settlements in Bengal and the Coromandel Coast subordinate to Fort St. George, thus creating the circumstances proper to declaring Madras a sovereign state toward the end of the seventeenth century.[52]
Yale's unsteady loyalty to company and charter, monarch, and religion
challenged an already vexed position the British had in a land of immense
wealth and power, no matter how many declarations were made in its behalf, or how many times its topography was renamed to reflect English
allegiance.[53] In 1695, the Directors' efforts to destroy Yale's character
through its suits and countersuits had the following unpleasant effect. The
king gathered a Privy Council consisting of his principal ministers and earls,
Prince George of Denmark and the Archbishop of Canterbury in order to
hear "the Case of Elihu Yale, late President of Fort St. George in East India."
The Privy Council then ordered the attorney general and solicitor general to
"peruse the Charter granted to the East India Company and report their
opinion as to the Company's powers in matters of Judicature and Courts in
the East Indies."[54] Thus a Parliamentary investigation of the juridical
powers the Company wielded in India was ordered, and the Company was
subsequently found guilty of a Pandora's box of dubious activities including

bribery and embezzlement, the end result of which was the destruction of their monopoly.[55]

Yale was deposed in 1692. Seven years later he returned to England and word of his immense fortune spread across the Atlantic to his native New England. In 1718, Cotton Mather, then the representative of the Collegiate School of Connecticut, asked the now aging Yale for help with that struggling institution. Yale responded by giving Mather a portion of the cargo he brought back from Madras. In addition to the books and textiles of more popular fame, this cargo included "31 pieces of Madras chintz, 2 pieces of choice cloth flowered with silver, a piece of rich satin with gold flowers, 13 pairs of large gingham sheets, two jars of mango chutney, a case of soy sauce" and "a jar of nutmeggs."[56] This gift resulted in a substantial endowment for the college, which then changed its name to honor its benefactor and the rest, as they say, is history.

One of the premises of this chapter is to think about the interlocking scales of historical representation: local, regional, national, transnational, and global. It is with that in mind that I refer to both the histories of nutmeg and Yale as events: something that happens that then becomes an occasion for other things to happen. Yale's seemingly random collection of things, acquired despite a history of administrative failure and donated to the college, result in the manufacture of a lasting name, an authoritative institution, and eventually, a new national space as the college Yale established to train Congregationalist ministers became, in the nineteenth century, a custodian of the American nation-state. Reading Yale's story as a history of American or British entrepreneurism must disconnect one life from another and focus on the lack of national coherence. But putting Yale's story next to the story of nutmeg refashions it as a transnational event.

Just as nutmegs shifted their cultural meaning from exotic spice to botanical lore, that is, eighteenth-century scientists abstracted them from their material significance to become a knowledge claim, so Yale's material life became abstracted to an institution of knowledge. James C. Scott accounts for this process of abstraction as a way of rendering landscapes legible. He claims, "certain forms of knowledge and control require a narrowing of vision" that, in turn, bring into "sharp focus certain limited aspects of an otherwise far more complex and unwieldy reality." Citing the example of forestry as it was developed in the early modern German European state, he emphasizes the distance that the administrative state has from various productive locales and the consequent distance between the kind of knowledge produced by the state—the standardization of measurement that, in a Foucauldian sense, putatively serves the interests of the modern subject—and local knowledge and practice.[57] Bruno Latour argues that the concept of modernity is a fantasy; that the state-created subject is always bumping up against the material reality of local differences. In the case of Yale, his refusal to administer Madras according to the charter created by the Right Honorourable Company in London, his propensity to engage in local trade and customs, to endorse a kind of miscegenation, fraternizing as he did with a Jewish woman and producing, not unlike those bastard hybrid nutmegs, his own hybrid progeny who stood to inherit the bulk of his and his brother's estate, provided too great a material obstacle for the Company to continue its business and thus he was deposed.

Driven by commercial interests, to be sure, what came back to Europe and England and the state of Connecticut in the ships from the Indian Ocean was totally unexpected and therefore introduced a new history. Yale's "jar of nutmeggs" turns into a university; nutmegs become the material metaphor for Yale himself that abstracts into an institution for furnishing authenticated knowledge: the desire for nutmegs turns into a knowledge of them.

It turns out that Connecticut is known as the "nutmeg state," and that one nickname for people from Connecticut is "nutmeggers," effectively associating, albeit affectionately, its people not with the majesty of statehood but with the commodity that has an infamous history. Connecticut's state motto also alludes to nutmeg's history. "*Qui Transtulit Sustinet* (He Who Transplants Still Sustains)" originates from Puritan beliefs in a sustained faith that survives being uprooted from native England and transplanted in New England. Yale College was created to teach this Puritan belief. But such a motto is also informed by other, more agricultural forms of uprooting and transplanting that defined not only early modern American life and its devotion to a providential Christian sensibility but the ways in which botanical experimentation reproduced the knowledge—hybrid or otherwise—that Pierre Poivre generated with his nutmegs on Mauritius. I don't think this is simply a happy coincidence, but I also don't think that there's a one-to-one correspondence with Yale's jar. Rather, the state's nickname came about, so the story goes, by the reputation Yankee peddlers had up and down the Atlantic seaboard for manufacturing false nutmegs out of wood and selling them to people anxious to prove their cosmopolitan tastes. Grating the shavings of wood or nutmegs into their pumpkin pies apparently didn't make much of a difference to these intrepid souls, but it didn't have to because the tastes were equally peculiar and neither commodity cured the plague. Authenticity, therefore, was irrelevant. What was relevant were the fantasies they produced. Yale turned product into fantasy; and what gets invented in this shift is the move from mercantilism to capitalism. Emerging from the unknowable void that separates Yale University, now itself separated from its Puritan origins, from Fort St. George is the mystified product, luxuries that become signifiers of exchange value, not use value.

National histories operate as the master text, as the institution of knowledge: what gets represented is the enduring sense of a unified, coherent state that progresses from its infancy, of course, but progresses nevertheless. One needs them in order to render India and Connecticut radically disconnected. Transnational stories function as a form of backward reading, a hysterical demand that continually desires nutmegs, real or fake. Putting the textual remainders of Yale's life beside the nutmeg furnishes a differently situated knowledge.

Transnational histories allow for discontinuities. No matter how much attention Yale University paid to the highly publicized IncredibleInda@60 event in October 2007, their investment in maintaining relations with India obviates the radical connection between the two disparate standpoints and renders the modern nation-state of India as part of Yale's obligation.[58] And so we move from the Master of Madras to Yale University.

I opened this chapter with a discussion of mortar; I am ending it by offering the material phenomena of mortar, walls, and nutmeg as embodiments of different kinds of spaces that in turn provide different kinds of histories. Richard White's account of scale—the social and historical categories superimposed upon "the past that the people of the past neither knew nor used"—is not simply a call to substitute one kind of history with another.[59] Rather, he argues that scales—global, transnational, national, and local—

are themselves largely spatial. It follows, then, that historicizing spaces may identify the ways in which they are socially produced. The spaces I have written about—Madras, London, and New Haven—are all ones that have been crucial to sustaining national hegemony in one way or another. It's not difficult to imagine how any of these cities contributed to the history of British imperialism or the American academy. But breaking them up into smaller spaces—White Town, Black Town, the East India House on Leadenhall Street, the Collegiate School of Connecticut, and the diminutive physical space occupied by nutmegs in cargo holds on ships—identifies different topographies that aren't intuitively connected either to national or imperial hegemony.

Walls, mortar, and nutmegs tell different stories that compete with national accounts of imperialism. The walls of Fort St. George and the wall separating White Town from Black Town performed the ideological function of maintaining the difference between Christians and "Gentues," but clearly that difference was one that had to be manufactured because it wasn't essential or visibly obvious. The initial flimsy structural design in some ways functioned as a metonym for the uncertain footing Britons had in Madras. The walls and the Indian mortar that solidified them disconnected one space from another and were valued because they created imperial space. In a sense, interlopers were critical to empire-building because their ideological transgressions were subsequently read as material ones. Soldiers slipping over the wall to consort with Black Town natives, Thomas Pitt and Jacques de Paivia making tremendous fortunes as private traders, or Yale violating Company directives and lining his pockets as well as William Langhorn who spent his tenure as governor trying to keep the walls intact—all these figures helped to detach White Town from Black Town even if these spaces were materially the same. Madrasi mortar compacted the space of early British imperialism, rendering buildings like the Fort and the Garden House— structures that embodied British sovereignty by the turn of the century—solid, sound, and stationary. Although this mortar was specific to its site—the local oyster shells that were broken and pounded to create a particularly sticky substance—the mortar moved. From the Fort and the Garden House to the East India House on Leadenhall Street to the Royal Society archive in Crane Court, from Black Town to White Town to London, the method of making the best mortar in Madras eventually found its way into the *Philosophical Transactions* of the Royal Society just as it continued to support the foundations of White Town. Mortar thus became a technology of empire but one whose specific genealogy had to be erased and replaced by the belief in the fallacious but necessary attribution to national ingenuity that Jemima Kindersley helped disseminate: Madras was built "entirely by the English."

Nutmegs offer a different challenge. They are valuable because they are able to move between spaces rather than become space, as mortar does. Occupying little space themselves—a pocketful of nutmegs was enough to support a man for life in the seventeenth century and to buy him "a gabled dwelling in Holborn and a servant to attend to his needs"[60]—they moved easily among the trade routes of the Indian Ocean. British interest in nutmegs was sparked, as Milton has argued, when Elizabethan doctors decided they cured the plague, but the major problem facing British entrepreneurs was locating their extraordinarily specific site, especially given their unfamiliarity with foreign and, to them, dangerous oceans. In a strange way, the precision of its geographical and topographical demands made nutmeg much more portable than not. Unlike mortar, which had the capacity to make things stay put (even if it managed to move along the same trade routes), nutmeg, seemingly locked into the microclimate and volcanic soil of the Banda Islands, had a very mobile history. Yale didn't need to make that perilous voyage: the nutmegs came to him via interlopers like Thomas Pitt. Coming back from Madras to the warehouses in England to the Port of New Haven, Yale's jar of nutmegs moved from being Puritan (the Collegiate School of Connecticut) to Yankee (the peddlers of false nutmegs), from being commodities traded within the closed spice trade circuits of the Indian Ocean to the Atlantic, to being signifiers of British health and wealth and later of the American nation-state.

THREE

Ice and the Production of British Climate

> "Butler!" yelled Ellis, and as the butler appeared, "go and wake that
> bloody *chokra* up!"
>
> "Yes, master."
>
> "And butler!"
>
> "Yes, master?"
>
> "How much ice have we got left?"
>
> "'Bout twenty pounds. Will only last to-day, I think. I find it very
> difficult to keep ice cool now."
>
> "Don't talk like that, damn you—'I find it very difficult!' Have you
> swallowed a dictionary? 'Please, master, can't keeping ice cool'—that's
> how you ought to talk. We shall have to sack this fellow if he gets to
> talk English too well. I can't stick servants who talk English. D'you
> hear, butler?"
>
> "Yes, master," said the butler, and retired.
>
> "God! No ice till Monday," Westfield said.
>
> —GEORGE ORWELL, *Burmese Days*

This conversation related in this chapter's epigraph takes place in the
small fictional British outpost of Kyauktada in Burma of the 1930s, then
part of British India. Orwell's stunningly cynical novel of the last days of
the British Raj dwells almost lovingly on details of discomfort. The prickly
heat torturing British skins, the uncomfortable cane chairs of the Club that
stamp their denizens with permanent "pine-apple" backsides, the stifling,
intolerable heat that overpowers even the most intrepid tennis player all
result in a constant and ill-tempered imbibing of lukewarm lime juice, gin,
and beer starting shortly before breakfast. Orwell's representation focuses

on maladies created by the local weather to which the British were particularly, if not exclusively, vulnerable. In an especially poignant articulation of the imperial prestige this vulnerability confers upon its subjects, the two "derelict Eurasians," Francis and Samuel, discuss the "torments" they suffer from prickly heat and the measures they take to avoid sunstroke with the *pukkha* characters Flory and Elizabeth, desperate for any acknowledgment of racial solidarity: "For the natives, all well, their skulls are adamant. But for us sunstroke ever menaces."[1] But unlike the early residents of Fort St. George, who solved the problem of inclement heat by adapting to local customs even if they eventually acquiesced to orders forcing them to wear English apparel, the occupants of Kyauktada willfully continue British conventions defying environmental logic.

The theory of tropical heat-related diseases inflicted on the skin has a long history of representation in the colonial literature of Great Britain, and Orwell is exploiting this largely ideological position in order to make a point about the relationship between climate, race, and imperial identity.[2] Clearly the climate of the Club house, described as "teak-walled" and "smelling of earth-oil," is hardly a cool sanctuary in which the British can take refuge, and the possibility of "no ice till Monday" is one that invokes a divine supplication more heartfelt than ones voiced in the other bastion of Britishness, the church. The mildew, the beetles, the *punkahs* blowing dust about—all of these images are tropes for a decaying empire that come home to roost in the discussion Flory and the Indian doctor have about the British Empire as "an aged female patient of the doctor's."

The heat doesn't seem to affect the native Burmese of Kyauktada. But it doesn't affect the Indian servants either who have been stationed there along with their British masters. In fact the butler thrives in these conditions, speaking English with a fluency that enrages his interlocutor who accuses him of having swallowed a "dictionary." It is almost as if the melting ice releases the butler's facility with English thus posing a threat to his brutish master far worse than the prospect of no ice after that day.

This threat is not idle. Orwell's novel, set in the waning days of the British Raj, is quite explicit about the demise of British imperialism and the rise of the native colonial administration left as its legacy. What is more interesting is the relationship of ice to colonial politics. The consolation of ice is most acutely felt as a means of alleviating the heat, and this novel makes no secret that the promise of immediate refreshment is anticipated with an eagerness bordering on anxiety, its absence contemplated with dread. What is less obvious is how ice acts as a metonym of British climate, and, therefore, as a signifier of cultivated thought. Veraswami, the Indian doctor who is "fanatically loyal" to British rule, demonstrates his devotion to the West and "honourable English gentlemen" by filling the shelves of his library with books of essays "of the Emerson-Carlyle-Stevenson type" but also by having a "big tin ice-chest" in which he keeps "whisky, beer, vermouth and other European liquors." The shelves at the Club can hardly compete with Veraswami's: in addition to outdated copies of *Punch*, *Pink'un* and *La Vie Parisienne*, the "forlorn 'library'" consists of "five hundred mildewed novels." Club members profoundly mistrust any conversation elevated above Mr. Lackersteen's boorish comments ("'How about that for a pair of legs . . . You know French, Flory; what's that mean underneath?'") and dismiss intellectual exchange as either "Bolshie" (Flory) or "literary diarrhoea" (Macgregor). Quarrels at the Club are much more successfully quelled by calling for another drink "before the ice goes." But the Burmese sun is powerful, and any solid British authority signified by the ice dissolves under its beams leaving them with "flat, clear drops of sweat" gathered on their faces and "damp patch[es] growing larger and larger" on the backs of their suits. Melted ice signifies the dissolution of British administrative control in this novel, and part of what fuels its characters' xenophobic rage is the knowledge that the real work of the empire is in the hands of their subalterns.

Ice works as a successful trope for Orwell's project, nicely positioning British colonial administrators, softened by years of swilling whiskey, in states of despair at the prospect of no ice until Monday. Ice has a metonymic relation to the metropole, functioning as the material that divides the "jungle life" of teak extraction from the urban pleasures of Rangoon, crowned by "the dinner at Anderson's with beefsteaks and butter that had travelled eight thousand miles on ice."[3] Grown corpulent under a sun that never sets, these characters lament the loss of the "dear dead days when the British Raj *was* the British Raj and please give the bearer fifteen lashes," hopelessly mired in a suffocating combination of nostalgia for rigid British rule (ice) and the realities of the present "malaise . . . of the long, deadly hours that were coming" (heat).

But what if ice is not only a trope? Certainly in Srinivas Aravamudan's useful account of the trope, the "swerve from self-adequation to surplus" seems at work here. Operating as a transitive, tropes of climate control—the *punkah* that stirs the air, the ice that cools the drinks—preserve the separation between the British Ellis and the Dravidian butler who, no matter how many English dictionaries he swallows, can never be confused for a Briton. The exchange Ellis has with the butler about the availability of ice occurs because the *chokra* in charge of operating the *punkah* has fallen asleep, leaving everyone's shirts stuck to their backs with "the first sweat of the day." Thus the metropole remains distinctly marked by British bodies suffering from the native climate. But as a *material*, ice erases this difference. The butler's inability to keep the ice cool and therefore intact ironically results in something other than his fear of punishment from the master. Rather, the butler's reasoned calculations and linguistic command—two characteristics of British administrative control—are the most obvious effect of melting ice, which sparks an overdetermined rage from Ellis who, far from being able to give him "fifteen lashes," is left muttering impotent complaints to his colleagues. Ice may be counted as one of the humble materials that make life a bit more bearable in heat, but it may also have a far greater role in the process of empire building than many have acknowledged.

In order to address the different connections ice has to empire, I want to consider two stories about ice, both occurring centuries before the one in which Orwell set his novel, in the very early days of Britain's long association with India. Benigné Poissenot described his encounter with a "cave glacière" in his *Nouvelles histoires tragiques* (1586). Passing through the Jura region of France in June 1584, he stopped at the city of Besançon and was offered the local wine chilled with ice for his refreshment. Despite the fact that the effects of radical climate changes characterizing sixteenth-century Europe were common knowledge—effects that included unusually long winters and prolonged periods of cold that extended well into the spring and summer months—Poissenot found the idea of an endless supply of ice during the summer something truly marvelous, particularly for luxurious purposes. He longed to see for himself the sources of this wondrous store of ice, and was taken to the Froidière de Chaux, through the dense woods on twisting paths to the opening of the enormous cave, "a place so terrifying that he was reminded of what is said to be St. Patrick's hole in Hibernia." Nevertheless he drew his sword and entered the cavern "all paved with ice, and with crystal-clear water, colder than that of mount Arcadia Nonacris." His reaction to the cave, despite his desire to see it, was primarily of terror:

> Whenever I looked upwards my whole body shuddered with fear and my hair stood up on my head, seeing all the upper part of the cave covered with great blocks of ice, the least of which falling on me would have been enough to dash out my brains and tear me in pieces: so much so that I was like the criminal whose punishment in Hell was to have a big stone continually threatening to fall on him.

Not an isolated historical instance, successive travelers stumbled across this natural refrigerator that continually supplied the people of Besançon with ice to cool their wine cellars. In fact, the Froidière de Chaux was steadily exploited for its ice until a disastrous flood in 1910 appeared to dissolve its reservoirs and ice never formed again.[4]

Some 200 years later, Sir Robert Barker, Fellow of the Royal Society, adventurer, observer, and early modern climatologist, had a similar experience in Allahabad, where he was regaled with sherbets and iced creams when "the thermometer stood at 112 degrees."[5] Barker's reaction to the availability of ice during the extreme heat of summer was far different from Poissenot's. Unlike the well-known "cave glacière" shown to Poissenot, Barker had "never heard of any persons having discovered natural ice in pools or cisterns or in any waters collected in the roads." In a letter to the Royal Society, he begged "permission to present to you with the method by which [ice making] is performed at Allahabad, Mootegil, and Calcutta." He describes the process by which porous clay vessels of water, buried in pits filled with "sugar-cane, or the stems of the large Indian corn" (whose "spungy nature . . . appears well calculated to give a passage under the pans to the cold air"), salt, and saltpetre, came to provide ice for these cities.

> From these circumstances it appears, that water, by being placed in a situation free from receiving heat from other bodies, and exposed in large surfaces to the air, may be brought to freeze when the temperature of the atmosphere is

some degrees above the freezing point on the scale of FAHRENHEIT'S thermometer; and by being collected and amassed into a large body, it is thus preserved, and rendered fit for freezing other fluids, during the severe heats of the summer season.[6]

Unlike Poissenot's cautious approach to the caves of the Jura, Barker was eager to observe foreign practices of ice making, and while he muses that "during [his] residence in that quarter of the globe, [he] never saw any natural ice," somehow this absence of a "natural" explanation seemed perfectly sensible to him. Although Poissenot expresses himself to be "burning with desire" to be shown the "great stalactites of ice" in the Froidière de Chaux, he interprets the actual sight of this cave and its natural source of ice both as a threat and a form of criminal trespass.[7] Barker, whose own eyes and ears have not seen or heard of a natural source of ice in Allahabad or Mootegil or Calcutta, nevertheless assumes that Indian techne could produce quantities of ice from no enduring source of chill. If one grants Poissenot the experiences a well-traveled man would have had in sixteenth-century Europe, then his reaction to the Froidière de Chaux is surprising. Glacial advances had left their infamous mark of famines, frosts, and floods on towns from Chamonix to Besançon. This area is described as

"a poor country of barren mountains never free of glaciers and frosts . . . half the year there is no sun . . . the corn is gathered in the snow . . . and is so moldy it has to be heated in the oven." Even animals were said to refuse bread made from Chamonix wheat. The community was so poor that "no attorneys or lawyers [were] to be had." Avalanches, caused by low temperatures and deep snowfall, were a constant hazard. In 1575–76 conditions were so bad that a visiting farm laborer described the village as "a place covered with glaciers . . . often the fields were entirely swept away and the wheat blown into the woods and on to the glaciers."[8]

Navigating his way through this area, Poissenot must have been familiar with the ways in which giant rivers of ice had radically altered the landscape, making natural phenomena like the Froidière de Chaux a rule rather than an exception. Yet Poissenot's terror of the "càve glacière" borders on the sublime, his visit there a crime punishable by hellish forces, a horrific prospect that, far from cooling his "burning" desire, invokes flames hotter than any contrived of human device.

By contrast, Barker seems eager to embrace and consume the results of "Asiatic" science, regardless of whether or not his own sensory perceptions allow him to believe in the manufacture and production of ice. Rather than resorting to supernatural explanations, he uses his own instruments of measure to render the results of an alien techne readable to him: "Upon applying the bulb of a thermometer to one of these pieces of ice, thus frozen, the quicksilver has been known to sink two or three degrees." Thus, from "an atmosphere too mild to produce natural ice, ice shall be formed, collected, and a cold accumulated, that shall cause the quicksilver to fall even below the freezing point."

The "promising advantages" he identifies make profitable "a short duration of cold [to] alleviate the intense heats of the summer season, which in some parts of India, would scarce be supportable, but by the assistance of this and many other inventions." Such profit, stimulated by the "Asiatic, whose principal study is the luxuries of life," could capitalize for the benefit of European visitors.

Barker is making clear connections between two forms of technologies, the "principal study" of Asiatic science, luxury, and the measurement of instruments of Western science, the thermometer, in order to determine that what he is seeing is, indeed, a collection of ice, despite the lack of any natural evidence. Poissenot's wonder at the availability of supplies of ice during the height of summer generates a "burning desire" to see such a phenomenon with his own eyes. Yet the sight of the "càve glacière" is one that produces an unreasonable terror, despite the abundance of natural evidence, anecdotal or otherwise, that makes this site perfectly natural. He resorts to the supernatural, trembling with fear for the hellish punishment that awaits his transgression. What accounts for these radically different reactions to the same story: the unaccountable production of ice during a season where no ice should be made? Or, rather, what accounts for these different reactions to the appearance of ice in the summer months when both travelers were writing deep in the throes of the Little Ice Age? Or, finally, can we account for five hundred years of ice out of context as a single, definitive historical period?

One answer seems transparent enough: what separates Benigné Poissenot and Sir Robert Barker is one hundred years in which the meteorological conditions might well have changed. What also separates them are a couple of continents, which might have experienced the effects of the Little Ice Age differently. This period, identified by historians as a series of climatic shifts extending from 1300 to 1850, repeatedly subjected England and Europe to droughts, diseases, famines, and eternal winters of bone-chilling cold. European reaction to these climatic shifts varied during Poissenot's time, but the prevailing sentiment from 1560 to 1600 was that the disastrous results of these shifts were supernatural in origin.

> As climatic conditions deteriorated, a lethal mix of misfortunes descended on a growing European population. Crops failed and cattle perished by diseases caused by abnormal weather. Famine followed famine bringing epidemics, bread riots in their train and the general disorder wrought fear and distrust. Witchcraft accusations soared, as people accused their neighbors of fabricating bad weather. . . . Sixty-three women were burned to death as witches in the small town of Wisensteig in Germany in 1563 at a time of intense debate over the authority of God over the weather. Witch panics erupted periodically after the 1560s. Between 1580 and 1620, more than a 1,000 people were burned to death for witchcraft. . . . Witchcraft accusations reached a height in England and France in the severe weather years of 1587 and 1588. Almost invariably, a frenzy of prosecutions coincided with the coldest and most difficult years of the Little Ice Age, when people demanded the eradication of the witches they held responsible for their misfortunes.[9]

The social climate of suspicion and fear, spurred on by a belief in the divine authority over weather, inevitably led people to decide that radical weather changes were supernatural in origin. Poissenot's reaction to clear evidence of the disruption of the natural cycle of seasons makes sense, especially considering that this was pre-Cartesian France (at least at the time that Poissenot wrote *Nouvelles histoires tragiques*). He describes how his initial desire to see the cave is gratified by his hosts, but later when the phenomenon is revealed to him, he realizes that such desire is nothing if it is not sin: the sight of a world turned upside down, where vast quantities of ice exist regardless of the season, but exist dangerously, ready to strike down the transgressor whose desire is too "burning." The accounts of witches being burned with alarming frequency as a punishment for altering the natural progression of the seasons saturated travel narratives in late sixteenth-century France. Despite the abundance of natural evidence for the "càve glacière," Poissenot, an intrepid adventurer and writer, could not help but be reminded of his own part in this unnatural phenomenon, particularly because he was enjoying its luxurious ramifications: iced refreshment during his summer sojourn in Besançon in late June 1584.

Barker's experience is rather different. It is true that both travelers were writing during the historical period of the Little Ice Age; however, epistemological changes shifted from the primacy of theodicy and recourse to the supernatural to the discourse of science to make unfathomable natural phenomena readable. For example, the emergence of the experiment as a crucial part of scientific methodology seems to be, in part, responsible for Barker's sangfroid in the face of no physical evidence of natural sources of ice. Drawing on the Hobbes-Boyle controversies of the seventeenth century, Steven Shapin and Simon Schaffer argue that "solutions to the problem of knowledge are embedded within practical solutions to the problem of social order, and that different practical solutions to the problem of social order encapsulate contrasting practical solutions to the problem of knowledge."[10] In the case of these radically different responses to ice that Benigné Poissenot and Robert Barker record, while chronological, historical, and geographical differences may easily contribute to the differences of their accounts, clearly the problem of social order is one that determines their discrete responses to inexplicable natural phenomena. The failure of witch sacrifices to alleviate the problems of climatic changes during the Little Ice Age may have contributed to seeking other solutions—often attributed to the "natural" linearity of a progressivist history—to account for problems of the social order. Thus, rather than invoking the supernatural, the wrath of God, and the dominance of theodicy as ways of determining causal issues, the emergence of a system of accountability based on empiricism might have had an equally radical influence on climatic etiology. Enclosed within a geographical economy of suffering, the issues of luxury seemed a sinful sign of excess, ones that could, without care, be castigated as sorcery.

And yet, Sir Robert Barker's "observations" of ice making in Allahabad— if that's what they were—are steeped in a kind of witchery: "promising advantages" were to be had to make profitable this "short duration of cold." A "comparatively short season of cold weather"—far from freezing—would magically supply a long season of intense summer heat with means of alleviation. Barker's thermometrical tests determine that "from an atmosphere too mild to produce natural ice, ice shall be formed, collected, and a cold accumulated." Such conclusions rely on something other than an analytical model of reason: Barker seems to be engaged in an analogical reason that determines that ice produced from no enduring source of chill is perfectly natural and reasonable, an alchemical paradigm that fully believes in the transfer of base substance into a sublime abstraction. That is, though Barker's recourse to the Fahrenheit system of measurement seems in line with empirical methodologies of control, the fact is that there is no discernible source of "natural" ice to be had. Instead, Barker reasons, the ice that is "formed, collected, and a cold accumulated" is one that is analogous to modes of ice collection that characterized European experience.[11]

Barker falls back on an older form of reason that suits his context. His presentation to the Royal Society is the "method by which it was performed at Allahabad, Mootegil, and Calcutta," presumably by Indian ice-makers schooled in this practice and fueled by "Asiatic" study, whose principal interest is the "luxuries of life." Untrammeled by any fears of witchcraft or allegations of sorcery—Barker is, after all, working well within the Newtonian confines of Enlightenment science that putatively put such nonsense to rest—Barker nevertheless seizes upon a medieval paradigm of science with which to make such a phenomenon readable to him and to his audience at the Royal Society. Porous clay vessels containing water, buried in pits of reeds, salt, and saltpeter, miraculously yield enough ice to cool a "season" of "intense heats," which "in some parts of India would scarce be supportable." The value of this form of "Asiatic" techne for European colonists is inestimable, especially considering not only "the assistance of this" but of "many other inventions" that could render European habitation possible. The rudimentary elements of ice making—water, clay, reeds, salt, saltpeter— produce the invaluable means not only for personal refreshment but the luxury of climate control. In short, with careful husbandry, Indian technology could Europeanize India's climate. Moving from the base substance of this method to its priceless consequences, Barker recalls the paradigm of an alchemical transfer to make legible to his London audience the value of his own "empirical" observations.[12]

British ventures into the New World during the Little Ice Age were also faced with the quandary of the colonist: although "demystifying nature, displaying bodily strength, and using technology all became measures of colonial power," as Joyce Chaplin argues, the British were faced with the undeniable refutation of those very elements of mastery.[13] Even if climatic similarities rendered their experiments in Arctic navigation familiar, they were, nevertheless, confronted with the utter paucity of techne to deal with overwhelming natural ice.[14] Limited to Arctic exploration because of Iberian domination, the English found themselves bested by "uncivilized" Inuit nations largely because of their assumptions about technologies:

> it was one thing to have expected to meet Asian migrants passing through the north and another to meet people who lived there. Indeed, [Martin] Frobisher's first crew was astonished that they had an audience for their efforts in the Arctic, which Best reported was "habitable" by people and animals. The people lived in a veritable desert, one where agriculture itself was impossible—something incredible to Europeans. In such surroundings, established opinions about raw nature and its transformation made little sense. The English puzzled over how the Inuit managed to stay alive in their region, and how they so quickly adopted European technology. Lack of sustained contact with the natives of Newfoundland, before 1612, created a continued context for puzzlement, especially as the English tried to make sense of the increased presence of European technology among the Abenaki, Beothuk, Micmac, and Montagnais, without the increased presence of Europeans.[15]

Sixteenth-century English explorers, then, initially found the conditions of excessive frost and cold in the New World a source of comforting familiarity, something that was quickly dispelled when it became in increasingly clear to them that European bodily might, technological superiority, and the conquest of a fickle nature were only illusions, and that far from confronting an uncivilized "tribe," the Inuit defied every tenet of their colonial discourse, including the supremacy of European techne. Even if Frobisher's first crew experienced this cultural disjunction with necessary immediacy, the dissemination of the discrepancy between European fantasies of Inuit and Eskimo tribes and the bitter realities environmental conditions wrought was slow to spread. After their stinging expulsion from the Thirteen Colonies some 150 years later, however, the British sought different contexts in which to articulate measures of colonial power.[16]

Barker's experience in India, then, has a rich and complex environmental footprint that extends not only over several continents but also over several centuries. What's interesting about these two stories about ice is not so much their chronological and geographical distance, both of which are fairly easy to imagine, but the insistent continuity of certain ideas about climate control and certain fantasies that imbricate climatic change, history, colonial ideology, technology, and scientific and literary discourse.[17] As historians Emmanuel Ladurie, Brian Fagan, Joyce Chaplin, and Linda Colley agree, it seems as if history is shaped by diverse conditions including climate change that could well alter the ways in which we understand such mainstays of discursive continuity as colonialism. Rather than thinking solely about the administrative implications that occupy postcolonial discussions of British colonialism—and by administrative I am referring to ideologies of domination and conquest that have structured postcolonial inquiry—it may also be important to think about circumstances putatively out of human control. It is not that administrative structures of power are not central to understanding colonialism—they are indeed necessary and crucial to understanding ideological production. But issues such as the weather, the climatic conditions of early modern Europe that had a critical place in imperial history may contribute to usefully complicating the a priori agreement that European Enlightenment historical and technological progress led to an inevitable global dominion.[18]

About twenty years before Sir Robert Barker was engaged in his thermometric observations, reporting his findings to the Royal Society, Samuel Johnson had finally completed his *Dictionary of the English Language*. This project, undertaken in 1746 and occupying his time and psychological wherewithal for the next ten years, was in part motivated by competing academies in Italy and France who were compiling dictionaries of their own, but it was also spurred on by Johnson's own interest in taxonomy:

> When I thus inquired into the original of words, I resolved to likewise show my attention to things: to pierce deep into every science, to inquire the nature of every substance of which I inserted the name, to limit every idea by a definition strictly logical, and exhibit every production of art or nature in an accurate description, that my book might be in place of all other dictionaries whether appellative or technical.[19]

Even if he dismisses this ambition as the "dreams of a poet," Johnson was clearly provoked by the mass of correspondence coming from the British East India Company, members of which were charting their interactions with an alien land in the interests of epistemology. His gargantuan undertaking, made "little assistance of the learned, and without any patronage of the great," is recorded with a certain bitterness in his preface to the *Dictionary* in which he recounts, among other things, the thankless task of the lexicographer. Doomed to "toil at the lower employments of life," to be "rather driven by the fear of evil than attracted by the prospect of good; to be exposed to censure, without hope or praise, to be disgraced by miscarriage or punished by neglect," is the fate of these "unhappy mortals," among whom is the "writer of dictionaries."[20] Johnson's representation of lexicographical labor is curiously prescient of later nineteenth-century descriptions of the thankless toil of colonial laborers who, for the most part, work tirelessly for the good of indigenous populations and the glory of the British crown with equal loyalty.[21] Unlike Barker's excitement at discovering new methodologies to improve the circumstance of Britons in India—discoveries that submit to the reputed rigors of European scientific measurement—Johnson calls up the figure of the "slave of science" rather than its "pupil," a position "doomed only to remove rubbish and clear obstructions from the paths through which learning and genius press forward to conquest and glory." Like Daniel Gabriel Fahrenheit before him (1724), Johnson's place in scientific enterprise is to provide a system of measurement against which the purity or reliability of analogical experiments may be tested. Finding "our speech copious without order, and energetic without rules" while "adulterations were to be detected without a settled test of purity," Johnson applies himself to the perusal of our writers; and noting whatever might be of use to ascertain or illustrate any word or phrase accumulated in time the materials of a dictionary, which, by degrees, I reduced to method, establishing to myself in the progress of the work the rules as experience and analogy suggested to me; experience, which practice and observation were continually increasing; and analogy, which, though in some words obscure, were evident in others.[22]

Johnson's preface is driven by a deeply xenophobic invective against the problems with stabilizing and even freezing language and meaning. The trials that frequent translation, combined with the "jargon" or "mingled dialect" that "serves the traffickers on the Mediterranean and Indian coasts," pose to the "settled" test of purity of English diction are "as much superior to human resistance, as the revolutions of the sky, or the intumescence of the tide."[23] Thus Johnson's toil, superhuman in his effort to stem the insidiousness of a "natural tendency to degeneration," is "devoted . . . to the honour of my country, that we may no longer yield the palm of philology without a contest to the nations of the Continent."

But it is equally important as testimony to the dissociation of mercenary work to establish ways for new discovery. Thus the "slave of science" and "pioneer of literature" engage the European burden of science (as opposed to the Asian science of luxuries) and provide an "established principle of selection," whether to the "chief glory of every people . . . its authors," or to "the propagators of knowledge . . . and teachers of truth," in which "[Johnson's] labors light to the repositories of science, and add celebrity to Bacon, to Hooker, to Milton, and to Boyle."[24] Unaided by the "soft obscurities of retirement or under the shelter of academic bowers," Johnson's cold "sickness and sorrow" nevertheless establish him as the proper lexicographical laborer. Johnson thinly veils his contempt for the soft shelter of French and Italian academies, posing his "gloom of solitude" against their temperate climates for the cooperative production of lexicographical work, warmed by relatively generous patronage. He positions the relatively enervative efforts of effete cultures by his own self-representation as the eighteenth-century man of letters driven to work tirelessly, courageously, and fruitfully in an intellectual and meteorological climate zone that produced fantasies of individualist control. Johnson's *Dictionary* may thus be read not only as emerging from British taxonomic and epistemological drives, but by the ways in which Hippocratic notions of climatic determinism fueled and propelled fantasies of colonial enterprise.[25]

Johnson completed his lexicography in 1755 and was known thereafter as "Dictionary Johnson," a title that both elevated him to the status that his honorary MA from Oxford, included on the title page of the *Dictionary*, had already determined, and designated him as the hopeless literary drudge that he feared he might be. Nevertheless, such an appellation made his contribution resonate in both the world of letters and in the world of scientific discovery. Barker, writing twenty years later, would have been familiar with his notions of standardization.

But more interesting, Barker is far more eager to embrace encounters with otherness in their sublime and uncanny forms and, consciously or not, emulates the same kind of double reasoning in his letter to the Royal Society regarding ice making in Allahabad as Johnson does in his preface to *The Dictionary*. Johnson sets about collecting examples of diction "from writers before the Restoration, whose works I regard as the 'wells of English undefiled,' as the pure sources of genuine diction." Barker sets about determining the purity of the ice he sees collected in clay vessels, observing that "plain water will become so hard as to require a mallet and knife to break it," and thus concludes that "these pieces," defined by the "bulb of a thermometer [whose] quicksilver has been known to sink two or three degrees," are, in fact, "ice."[26] Both self-consciously announce when they recourse to analogical reason when analysis fails: when Johnson's "experience" fails, "analogy, which though in some words obscure, [is] evident in others" supplies the gap, while Barker, undisturbed by never having "heard of any persons having discovered natural ice in the pools or cisterns or in any waters collected," is willing to replace his lack of empirical evidence with Indian techne, even if he uses instruments of his own understanding to determine authenticity.

What allows for these figures to navigate back and forth between different paradigms of reason, given the primacy of the discourse of experiment: the "problem of generating and protecting knowledge" as a "problem in politics" and "the problem of political order always involv[ing] solutions to the problem of knowledge"?[27] Returning to the story of Benigné Poissenot and his response to the "càve glacière," one way of accounting for these two pioneers of science, Johnson and Barker, is to think about knowledge is in terms of climate control. Obviously, both were writing during the Little Ice Age, Johnson experiencing the rigors of climatic shifts perhaps a bit more keenly than Barker. But if we think about the ways in which Johnson's call for fixity is articulated in his invectives against translation, we can get an idea of the contrary drives that determine his enterprise. On the one hand, change is inevitable and sometimes even progressive: the "French language has visibly changed under the inspection of the Academy, the style of Amelot's translation of Father Paul is observed by Le Courayer to be 'un peu passé'; and no Italian will maintain that the diction of any modern writer is not perceptibly different from that of Boccace, Machiavel, or Caro." On the other, is the great "pest" of speech—the "frequency of translation"—that, according to Johnson threatens to destroy its integrity. "Let them," he enjoins, "stop the license of translators, whose idleness and ignorance, if it be suffered to proceed, will reduce us to babble a dialect of France."[28] Johnson's lexicographical contribution somewhat magically performs the impossible task of freezing meaning within the rigid confines of the "wells of English undefiled," willfully unaffected by the unavoidable movement of discourse from continent to island. His terror—if that's the right word—of fluid linguistic parameters slides into his own chilly indifference at the conclusion of his poignant preface, where he declares that "success and miscarriage are empty sounds" that he therefore dismisses with "frigid tranquility, having little to fear or hope from censure or praise."[29]

While writing a different text in a similar climatic context, Barker's motives are not entirely dissimilar. The fantasies about controlling seasons of excessive summer heat using the inventions and techne of Indian science circumvent the actual situation of these East India Company officers, stuck in an inhospitable landscape, and lured with the promise of fabulous profit. Regaled as he is with frozen "sherbets, creams, or whatever fluids are intended to be frozen," when the thermometer has registered above 100 degrees Fahrenheit, he can construct an entire cultural imaginary of a contained and controlled landscape, untrammeled by climatic realities. Untroubled, it would seem, by Johnson's fine distinctions between the "Mahometan" pursuit for "procuring the conveniences of life," a practice situated "a little above barbarity,"[30] Barker embraces this uncanny production of comfort and luxury and sees the possibilities for British profit, even if he embraces alien techne in order to put it to proper work. Clearly, being cold is associated with being English, no matter how it is produced or procured. Benigné Poissenot's reaction to the gelid realities of glaciers encroaching upon arable farmland, thus holding the very fragile agricultural existence accountable to the whims of weather, was that such climatic and topographical phenomena were the cruel realities of existence under the auspices of a distant and unpredictable god. This "knowledge" renders his terror of the "càve glacière" that much more reasonable. Such a natural formation deserved the kind of respectful trepidation with which he approached it, given the fact that the land and climate were, according to sixteenth-century knowledge, part of an entirely capricious supernatural will.

The end of the Little Ice Age—1850, according to Brian Fagan—wrought other kinds of shifts in the British relationship to Indian landscapes. More secure in their imperial purchase, Britons seemed to be confronted by the material realities of intense heats, and rather than turning to local technologies and local knowledge—which, by this point, they had dismissed as primitive, obscure, ineffectual, and rudimentary—they sought relief in the natural landscape. Kavita Philip documents the ways in which hill stations and hill settlements became part of the annual migration of Britons living in India. Referring to a popular ditty of the early twentieth century, Philip argues

> Not only is the Nilgiri climate a relief from that of the Indian plains . . . but it surpasses even that of England . . . the Niligiris are represented as paradise: as England on a grander scale, a land blessed with immortality. The first two stanzas make clear who these immortal residents are: it describes "the old fellows." They are, most likely, planters or foresters, sitting outside "the clubhouse"—a mandatory component of every British settlement, and a strategic cultural site for the reproduction of colonial ideologies . . . In the Nilgiris . . . the Englishman escapes from the tyranny of regulated, ordered temporality into a space that stands outside of time. Nilgiri sunshine has "the art" that English sunshine doesn't—this appeals to a view of art as outside sociality and temporality, eternal, unspoilt by the grimness of production relations. Like pure art, then, Nilgiri nature is unburdened by social functions; its manifestations are fairylike, full of flowers and birds.[31]

Having abandoned the real or imagined promise of deploying Indian technology to suit their domestic needs, nineteenth-century Britons took charge of the land. Their hill stations functioned as places from which they could escape both the rigors of seasonal heats and the rigors of administrative duties. As Philip argues:

> Officials in the hills could easily see themselves as invested with god-like qualities: a sense of omnipotence emerges, in this account, from the language of laying out, settling, and planning the lives and landscapes of the hills. Hill stations were relatively isolated from the state bureaucracy, and hence individual administrators were less accountable to higher officials. This increased personal power allowed hill administrators to feel like lords of their own kingdoms.[32]

The terms have shifted somewhat. Not just meteorological issues of chill and warmth, but topographical terms of height and depth become the new, more controllable markers for British power. Unlike Barker's tabula rasa of eighteenth-century Bengal, these colonial administrators found themselves, like Benigné Poissenot, bound once more to the structural realities of the landscape.[33] The advent of a more complicated political relationship to imperial rule may have resulted not only in a more layered bureaucracy, one that necessitated a change in the ideologies of imperialism that now envisioned British rule as social missionary work, but one that was also shaped by climatic shifts on the home front. To be sure, nineteenth-century Britain was still faced with the vagaries of weather—witness the legendary nineteenth-century yellow fogs of London under cover of which all sorts of potentially nefarious industrial deeds were performed—but nothing like what it had faced during the depths of the Little Ice Age. Conversely, the climatological perceptions of India were now being cast in other Hippocratian forms: the frightening heat of the summer seasons begat restrictions as insurmountable as Benigné Poissenot's reaction to the "càve glacière." Thus the hill stations managed to create British spaces that were eccentric to British administrative control, spaces that perhaps relied more and more on forms of local knowledge jealously guarded from the teeming hordes of Indian civil servants and reserved as exclusively British spaces. Again, as Philip suggests, the problems of administrative control were articulated in an endless bureaucracy:

Take this account of official duties in the hills: 'what a god-send and relief it is to get a little elbow-room, a little broad freehand work, a chance of action not absolutely determined by Codes and Manuals. . . . Oh, the relief of getting away [from the plains;] from niggling and pottering over Section X Subsection Z to a wider world where one lays out roads and reservoirs, plans for reserves and areas for cultivation, settles questions of rights and tenures, has, in a word, a chance of getting something done that shows.[34]

Showing one's work here capitalizes on another model of administration that Scott identifies as a form of mapping control over local areas from an abstracted governmental center regardless of its peculiarities.[35] This colonial administrator poses the problem as one that pits tedious bureaucratic abstraction—"Section X, Subsection Z"—against the material pleasures of road building and settlement planning that amount to another form of climate control. Controlling the land was a form of controlling the climate. Decisions British colonial administrators made regarding the ground shaped the climate, both in the Nilgiris and in other important hill stations in Bengal like Darjeeling.[36] Although the move from controlling meteorological conditions to controlling topographical ones seem different, they are linked by politics and power. Barker's plans for improving British habitation in eighteenth-century colonial Bengal by exploiting Indian techne were modified in the nineteenth century. They now emerged as the East India Company's decision to claim hills for their topographical control, in effect amounted to feeding the fantasies of climatic and political control.

In January 1883, the *New York Times* published an article from *Chamber's Journal* on ice making in India in which the author allegedly hails the advent of a technology that will decrease dependence on foreign ice.

When I came out in 1853, Calcutta, Madras, and Bombay were wholly dependent on American ice, supplied by the Tudor Ice Company, and retailed at two *annas* the *ser*; that is, two pounds of ice brought from America were sold in India for 3d.! . . . Science came to our aid; and sulphuric ether and ammoniac machines came gradually into vogue, and latterly Carré's wonderful pneumatic machine. . . . With these great appliances, block-ice is now available in districts where it could not formerly be had at from 11/2 to 2 annas per *ser.*

In an uncanny return of the French "càve glacière," Ferdinand Carré's invention of the compressor brought with it another fantasy of climate control for this author. What is striking about this article, however, is not so much the ways in which new forms of technology and machinery change the climatic conditions of British India, nor the curious dependence on American ice, but that most of the article is taken up with the author's fascination with "an industry peculiar to the cold weather, which, except in small stations, is fast dying out, and that is the manufacture of ice." While deploring the mercenary habits of the Americans and lauding the advent of European technology to obviate this humiliating dependence, the author nevertheless describes in great detail the same process of manufacturing "artificial ice" (on which "the *mofussil* [up country] was entirely dependent") that Barker had written about a hundred years earlier. Although the process is "tedious and expensive," it is saturated with nostalgia and the danger of its disappearance is preserved by this author in his account, sent back to London, published by the *Chambers Journal*, later taken up by the relatively new *New York Times*, to be disseminated to a transatlantic audience. Nevertheless, the American association with ice is not simply coincidental and therefore irrelevant to imperial history. This association makes possible unlikely transoceanic connections that complicate the story of British colonial enterprise. Frederic Tudor's Boston-based company started supplying ice to Calcutta in 1833. Although the journey was at least four months' duration from Massachusetts Bay, ships laden with 180 tons of ice in Boston entered the Hoogly River (a tributary of the Ganges) with 100 tons of ice still remaining for sale. Calcutta became Tudor's most rewarding interest, yielding over $220,000 in profits. The ice Tudor sold was from several ponds in Massachusetts but the most interesting one from a literary standpoint was Walden Pond in Concord. Thus the water most commonly associated with American transcendentalism was, when frozen and cut and harvested by Tudor's company—turned, in fact, into a hard commodity—a source of immense profit, hardly in keeping with the belief in a spiritual ideal untrammeled by physical or empirical realities preached by Emerson or Thoreau or even Carlyle. In a curious turn of the intellectual ideals articulated by the British in general and Johnson in particular, Thoreau's *Walden* drew from the Vedic scriptures, a well putatively undefiled by the rigid doctrine of established religions. American water from Walden Pond, inspiring at least one major indigenous intellectual movement, was thus simultaneously engaged in shaping the material realities of British colonial life in Bengal.

Thoreau had articulated this idea before, commenting on both its abstract and substantive possibilities. In 1854, he writes:

> In the morning I bathe my intellect in the stupendous and cosmogonal philosophy of the *Bhagavat Geeta*, since whose composition years of the gods have elapsed, and in comparison with which our modern world and its literature seem puny and trivial; and I doubt if that philosophy is not to be referred to a previous state of existence, so remote in its sublimity from our conceptions. I lay down the book and go to my well for water, and lo! there I meet the servant of the Brahmin, priest of Brahma, and Vishnu and Indra, who still sits in his temple on the Ganges reading the Vedas, or dwells at the root of a tree with his crust and water-jug. I meet his servant come to draw water for his master, and our buckets as it were grate together in the same well. The pure Walden water is mingled with the sacred water of the Ganges.[37]

Writing in his journal in the winter of 1846–1847, Thoreau reveals a much more concrete connection between the two bodies of water while he watched Tudor's ice-cutters carving blocks of ice for export:

> The parched inhabitants of Madras Bombay—Calcutta—Havana—Charleston & New Orleans drink at my well—While I incredulous read the vast cosmogonal philosophy of Ancient India—in modern New England The Brahmen's Stoic descendant still sits in his native temples and cools his parched lips with the ice of my Walden well. . . . If I am not a modern hindoo we are near neighbors— and by the miracle of commerce we quench our thirst and cool our lips at the same well. . . . And concord fixed air is carried in that ice to mingle with the sultry zephyrs of the Indus & the Ganges.[38]

Both texts address an imagined community of drinkers and both texts use the trope of water and ice to fashion such a transoceanic community. The air of Concord, "fixed" in ice, melts into the "zephyrs" of India in this fantasy of inspirational exchange, and yet it is not a mutually constitutive relationship but, rather, another articulation of Imlac's winds. This fantasy of communal exchange, however, is articulated only in his journals and notebooks; structurally, the global community of like-minded Unitarian transcendentalists extends only as far as Europe. The influential and self-taught Universalist, Orestes A. Brownson, imagines this community coming from

> every wind from all quarters,—from France, from Germany, from England even; and Carlyle, in his *Sartor Resartus*, seemed to lay his finger on the plague-spot of the age. Men had reached the centre of indifference; under a broiling sun in the Rue d'Enfer, had pronounced the everlasting "No." Were they never able to pronounce the everlasting "Yes"?[39]

Ventriloquizing Carlyle, Brownson comments on the enervated state of Unitarianism and the cry "from the bottom of their hearts for faith, for love, for union"[40] is one he imagines to be from "all quarters." This plea, articulated under the "broiling sun" of Diogenes Teufelsdröcke's encounter with the everlasting no, seeks the wintry paradise of Thoreau's *The Natural History of Massachusetts*. Thoreau writes that the

> doctrines of despair, of spiritual or political tyranny or servitude, were never taught by such as shared the serenity of nature. Surely good courage will not flag here on the Atlantic border, as long as we are flanked by the Fur Countries. There is enough in that sound to cheer one under any circumstances. The spruce, the hemlock, and the pine will not countenance despair. Methinks some creeds in vestries and churches do forget the hunter wrapped in furs by the Great Slave Lake, and that the Esquimaux sledges are drawn by dogs, and in the twilight of the northern night, the hunter does not give over to follow the seal and walrus on the ice.[41]

Thoreau's musings over expanding the horizon of common drinkers to like-minded thinkers, even if privately expressed, may reflect on his own state of confinement under Emerson's shadow.

Curiously, Dr. Veraswami's library does not include Thoreau's work or, at least, Orwell chooses not to mention Thoreau by name, only essays by the "Emerson-Carlyle-Stevenson" triad which, according to the doctor, have "a moral meaning." This omission may not have been deliberate, although it seems as if Orwell joins Emerson in his benign dismissal of Thoreau's contributions. Given Thoreau's propensity to embrace all sorts of drinkers at his Walden well including Brahmin servants, one wonders if the British Orwell, while critical of colonial xenophobia, is nevertheless part of the culture.[42]

The omission of Thoreau from the roster of literary transcendentalists thus stages a marginal resistance to New England's spirit of engagement, entrepreneurial or otherwise. After all, Massachusetts—once the center of a British colony, now signifying Britain's imperial failure—ends up supplying ice, to Britain's detriment, to their new imperial interest. Doctor Veraswami's "passionate admiration for the English" conveniently consolidates a united front against any American invasion, just as Oxbridge-trained Indian students would join their English chums in belittling "American" English.

Thoreau's sublimation of material ice into philosophical fantasies of unity would hardly have been popular with the waspish Ellis who articulates his real fear of black skins in tropes of liquidity, focusing particularly on the unlucky Doctor "Very-slimy": "Why else do you go to that oily little babu's house every morning, then? Sitting down at table with him as though he was a white man, and drinking out of glasses his filthy black lips have slobbered over—it makes me spew to think of it."[43] Ellis's vitriol removes him from the community of drinkers Thoreau imagines in his journal. The possibility of sharing drinks is embodied by the fear of touch, of contaminating "slobber" that places Ellis in a curious state of abjection, the "retching that thrusts me to the side and turns me away from defilement, sewage, and muck."[44] For Ellis, ice operates as a signal limit. Ice separates Club life—the "strategic cultural site" for the dissemination of colonial ideologies, as Philip reminds us—from the material realities of "comfortless camps," "sweltering offices," and "gloomy dak bungalows smelling of dust and earth-oil."[45] Its comforting edges are palpable; it slakes the intolerable thirst for dominion and creates an illusion, albeit briefly, of British climate.

Brian Fagan declares that today "our ecological sins seem to have overtaken our spiritual transgressions as the cause of climatic change."[46] Once at the mercy of a frozen landscape, of a supernatural wizardry that randomly determined the economic, social, and cultural wherewithal of a pre-Cartesian Europe, Enlightenment reason and colonial administration together focused attention away from meteorological conditions and toward a topographical control of land that shaped political power in crucial ways. Eighteenth-century interest in taxonomy, working in tandem with scientific discovery, sought ways to freeze the fluidity of exchange into extant meaning. Johnson's single-handed hardiness produced a tour-de-force that "soft"

Continental retirements and "obscure" academic bowers could not, and provided a reconfigured British colonial force with a lexicon of power.[47] Thus forms of Indian techne, far from being sources of either terror or derision, became possible solutions to anglicizing the weather. Administrative duty, however, forced Britons to pay attention to the land and the ways in which topography contributed to an increased presence of power. The fluid exchanges between ice-makers in India and Sir Robert Barker were then replaced by harder-edged boundaries that, nevertheless, wrought climate changes whose effects are still being felt. Fagan's "ecological sins," committed in the name of a technological progress, caution us to pay attention to the land, and, perhaps, the medieval retribution visited upon spiritual transgression. But whether or not we heed the hard realities of material landscapes or the conditions outside human management, our decisions are controlled by the desire to control that may, ultimately, shape our destinies.

FOUR

Inoculation and the Limits of British Imperialism

> Even the worst case of smallpox can be cured in three days. Dissolve one ounce of Watkins cream of tartar in a pint of water. Drink this at intervals when cold. It has cured thousands, never leaves a mark, never causes blindness, and avoids tedious lingering.
>
> —ADVERTISEMENT, *The Woman at Home* (1893)

> The State Journal had suggested in January, 1869, "a preventative medicine cream of tartar and sulphur," and advised: "It is worth trying, it costs little, and cannot fail to do much good. Take one ounce cream of tartar, 2 ounces flowers of sulphur, and mix well in a pint of molasses; dose, one teaspoon on going to bed (adult), ½ teaspoon (children). In ten days or two weeks it will have cleansed the system effectually."
>
> —EFFIE R. KNAPP, *Lane County Pioneer*, December 1961

In chapter 3, I mentioned a moment from Orwell's novel *Burmese Days* in which the protagonist, John Flory, recalls the pleasures of his youth and remembers dining at Anderson's restaurant in Rangoon where "beefsteaks and butter" have traveled thousands of miles packed in ice to grace the tables of this venerable institution. In a predominantly Hindu culture ruled by the belief in the holiness of cows, beefsteaks were rare, although butter—mostly in the form of clarified *ghee*—would be readily obtainable. The local availability of butter or beefsteaks, however, isn't the point. Part of what determines Anderson's attractiveness is the fact that these comestibles have journeyed from the west and from Britain in particular and are, therefore, metonyms of the metropole, ready to be consumed by suitable subjects. These foodstuffs supplied the appetites of British colonials anxious to claim solidarity with English culinary customs even if their daily diets were shaped by cultural hybridity.[1] But beef and butter weren't the only bovine products that were packed in ice and shipped to India in the nineteenth century. The smallpox vaccination campaign started in 1880 in Britain was waged with a peculiar intensity in British India, and one of the commodities regularly coming in on ships was the cowpox vaccine Edward Jenner had developed in England's west country. This in and of itself is not particularly noteworthy unless one is aware of the fact that smallpox inoculation had been practiced in India for hundreds of years before Jenner developed his vaccine. In this chapter I examine some of the ideological battles fought over the historical development of this techne that defined the limits of the British imperium.

Buried in the pages of an 1893 issue of the popular magazine, *The Woman at Home*, is a small advertisement for Watkins Cream of Tartar that offers a recipe for the cure of smallpox. Quackery was not uncommon in the Victorian era when medical practices were emerging and solidifying as scientific discourse, and the advertisements for quick fixes to dread diseases had a popular, if uninformed, following. In fact, the popularity of quackery tended to reflect, often with uncanny accuracy, the development of serious medical practices.[2] So it seems as if Watkins Cream of Tartar is merely continuing a marketing tradition set in mid-Victorian England. What is remarkable about this recipe is that it is directed toward this particular cure of smallpox. This disease had been endemic in Britain since the seventeenth century and had a long, unhappy history of periodic epidemics from 1628 (before people began charting deaths) to the most recent one in 1871. Moreover, there was an established method for preventing smallpox infection that had been in place for a number of decades.

Almost a century before this advertisement appeared, Edward Jenner developed the cowpox vaccine that he then used to inoculate James Phipp. The unimaginable success of this practice resulted in a series of laws in 1853 that made the smallpox vaccine mandatory among children. Even earlier, when there was more resistance to the idea of smallpox inoculation—the Royal College of Physicians in London was silent on the matter, the Paris Faculté de Mèdecine openly hostile—two London physicians successfully argued the case for smallpox inoculation to members of the Royal Society. John Arbuthnot (1665–1735) and James Jurin (1684–1750) tallied the number of deaths occurring from natural and inoculated smallpox and offered quantitative evidence that fewer people died from inoculated smallpox.[3] Quackery took many forms but it tended to reproduce, albeit falsely, the common methodologies that medical research had developed: the consumption of pills, for example, to ward off the effects of diseases or to cure contracted infections. So why, in the last decade of the century, during the most sophisticated period of medicine Britain had ever known, would this anachronistic recipe without the remotest connection to any true cure have even the small credence of an advertisement in a prominent magazine?

To answer this question one has to look at the etiology of inoculation, both as medical practice and philosophical concept. The earliest use in English (1589) refers to the grafting of plants. Not surprising, then, Lady Mary Wortley Montagu, writing to Sarah Chiswell in April 1717 from Adrianople (Edirne), uses this horticultural reference to define what she observed in Turkey: "Apropos of distempers, I am going to tell you a thing that I am sure will make you wish yourself here. The smallpox, so fatal and so general amongst us, is here entirely harmless by the invention of engrafting (which is the term they give it)."[4]

It is unclear whether or not the process Montagu represents was termed "engrafting" in Turkish or whether she was translating a term that had yet to exist in English, but what is clear is the emphasis Montagu places on "here": the town of Adrianople, a place, she assures her interlocutor, that is hardly composed of the "solitude you fancy me" but, rather, is an enviable location primarily because of its cultural sophistication. Indeed, she adds, "the French ambassador says pleasantly that they take the smallpox here by way of diversion as they take waters in other countries," offering his testimony to the refinement of the practice.

Describing the process, Montagu takes pains to emphasize the fact that it is predominantly a female practice:

> There is a set of old women who make it their business to perform the operation. Every autumn, in the month of September, when the great heat is abated, people send to one another to know if any of their family has a mind to have the smallpox. They make parties for this purpose, and when they are met (commonly fifteen or sixteen together) the old woman comes with a nutshell full of the matter of the best sort of smallpox and asks what veins you please to have opened.[5]

She also writes of her intent to "try it on my dear little son,"[6] and adds:

> I am patriot enough to take pains to bring this useful invention into fashion in
> England, and I should not fail to write to some of our doctors very particularly
> about it if I knew any one of 'em that I thought had virtue enough to destroy
> such a considerable branch of their revenue for the good of mankind, but that
> distemper is too beneficial to them not to expose all their resentment the hardy
> wight that should undertake to put an end to it.[7]

Far from championing English medical practice over the dubious practices
of Turkish women in *souks*, Montagu's cynical representation of the British
medical profession clearly positions avaricious doctors as obstructions to any
kind of innovation, foreign or otherwise.[8] Andrea Rusnock accounts for
some of the deep-seated antipathy to the practice, arguing that the homeo-
pathic methodology was counterintuitive to the Hippocratic maxim. She
also argues:

> Because of its non-European origins, some such as the physician William
> Wagstaffe, author of the influential pamphlet *A Letter to Dr. Freind Shewing the
> Danger and Uncertainty of Inoculating the Small Pox* (1722), disparaged inocula-
> tion because it was practiced "by a few *ignorant women*, amongst an illiterate
> and unthinking People." And finally, English nationalists raised Hippocratic
> objections to a practice developed in a foreign land (Turkey) for a foreign
> people: It could not possibly suit the needs of the Christian, meat-eating,
> English.

Montagu's remarks, however, challenge such logic. In the same letter to
Sarah Chiswell, she dismisses the "Grecian" (Christian) method—one that
marks the sign of the cross on the body instead of opening larger veins—as
"superstition," noting that this method "has a very ill effect." Her comments
about British doctors, then, may also be closely linked to issues of creating
and opening a feminine discursive space, something that spurs her to "have
courage to war with 'em," and entreat Sarah Chiswell to "admire the hero-
ism in the heart of [her] friend."[9] However critical Montagu is of the profession, she refrains from an explicit critique of their national identity; by
contrast, her patriotism, with which she characterizes her intentions, enables
her to seek foreign techne in order to counteract smallpox. Montagu's campaign did in fact trigger important experiments with smallpox inoculation,

including the public injection of six Newgate prisoners in August 1721, whose recovery from the disease induced by this practice in the following month encouraged Princess Caroline to inoculate members of her own family.[10]

Countering Montagu's defense of Turkish techne, Srinivas Aravamudan reads her peregrinations as an example of levantinization, arguing that her turn homeward in the penultimate letter to Abbé Conti and her introduction of inoculation in England together function as a critically renegotiated anglophilia.[11] Certainly the entry of the "Levant" into English cultural consciousness—incorporated as an English term when English merchants negotiated for shipping rights with the Grand Turk in 1579—may have been perceived as a less insidious, milder capitulation to the threat of a darker, more complete fall into alterity.[12] Such "falls" occurred with an alarming frequency in the early days of Fort St. George and the first governors of Madras were challenged almost daily to keep their English factors English. Montagu's own "fallen" son, Edward Wortley Montagu Jr., claimed to be Turkish, and spent his life in a vexed relationship with England, claiming to be Muslim to his death in 1776.[13] Defoe's reference to the Levant in the opening of *A Journal of the Plague Year* may be read as similarly apotropaic. Linking Levantine commerce with the entry of the plague into the European body, Defoe makes a clear connection between trade and disease, and even if conveniently displaced onto Holland, his insouciant comment about the unimportance of the disease's etiology suggests a degree of tolerance toward Levantine "goods":

> It was about the beginning of September, 1664, that I, among the rest of my neighbours, heard in ordinary discourse that the plague was returned again in Holland; for it had been very violent there, and particularly at Amsterdam and Rotterdam, in the year 1663, whither, they say, it was brought, some said from Italy, others from the Levant, among some goods which were brought home by their Turkey fleet; others said it was brought from Candia; others from Cyprus. It mattered not from whence it came; but all agreed it was come into Holland again.[14]

Like the gossips Montagu observed—the "ignorant women" William Wagstaffe scornfully dismissed—infection, for Defoe, travels discursively. While Aravamudan argues that transported in Montagu's letters about her

Turkish travels is a form of inoculation that strengthens the British corpus against foreign invasion, Defoe's fictionalized account of 1665 imagines a virulent disease transmitted "from the letters of merchants and others who corresponded abroad" to the neighborhoods of London with devastating effect.[15] In this case, levantinization falls short of its tropicopolitan duty, but is this a paradigmatic failure or the problem of a xenophobic archive?

It turns out the etiology of inoculation extends beyond Montagu's discoveries in early eighteenth-century Constantinople, beyond the introduction of the practice in the *Philosophical Transactions* of the Royal Society, the Royal College of Physicians and beyond Jenner's injections. Montagu's letters, eagerly read by recipients who composed the brightest of the English literati, and her appeal to the medical profession—who were largely aristocratic—for better or for worse, were all predicated on the idea of aristocratic privilege and British pre-eminence.[16] European fantasies about the munificence of Levantine luxury have, as many have pointed out, structured the ways in which they imagined the spaces of political power: the courts of the Grand Signior, the harem, and the hammam. Montagu's letters[17] have been variously read as correctives to English travel writers, where she uncovers a highly politicized feminine discursive space with her entry into the harem and hammam, or as a replicated masculine gaze into forbidden territory, inflected by feminine orientalism.[18] Whatever the case, her epistolary accounts are offered to an audience defined by the court. Dismissing English travel writers as "common," Montagu reinforces the notion that authentic travel writing is the privilege of ambassadors who bring back new strategies for negotiating and countering alterity, not the labor of merchants who are more prone to spread disease and misrepresentation along with their wares.

However, other voyagers routinely ventured even farther east than the Levant. Working under vastly different circumstances than Lady Mary Wortley Montagu and her retinue, British East India Company "servants" contributed an impressive arsenal of foreign techne to British archives in the massive correspondence between the Court of Directors in London and their soldiers, factors, merchants, agents, writers, and ministers. These often underpaid, overworked employees, thoroughly taxed by their subordinate relation to powerful Mughal emperors monarchs, had reported from the late seventeenth century on of the practice of smallpox inoculation in Bengal, but it wasn't until the eighteenth century that these reports were formalized and published in the *Philosophical Transactions*. Robert Coult, for example, writes from Calcutta in 1731:

> Here follows an account of the operation of Innoculation of the Smallpox as performed here in Bengall: taken from the concurring account of Several Brahmans and physicians of this Part of India. The Operation of Innoculation Called by the Natives Tikah has been known in the Kingdom of Bengall as near as I can learn about 150 years and according to the Brahamanian Record was first performed by one Dununtary a Physician of Champanagar, a small town by the Sydes of the Ganges about halfway to Cossimbazaar whose memory is now holden in Great Esteem as being thought the Author of this Operation, which secret, say they, he had Immediately of God in a Dream.[19]

Coult's documentation of the operation *tikah* (the Hindi word for spot) notes the "concurring account of several Brahmans and physicians" even as he conflates medical methodology with a reference to a mystified dream narrative. What is interesting is the similarity between the wandering groups of Brahmin healers spreading medical knowledge as well as dispensing inoculations against smallpox (among other things) and the rambling bands of East India Company servants eager to contribute to the storehouse of scientific knowledge embodied by the Royal Society. Zaheer Baber identifies the origins of the social organization of medicine in medieval India. During the late Vedic period (1500–500 BC), the cooperation between carpenters, healers, and priests that constituted medical practitioners were denigrated as impure "because of their constant bodily contact with people of various status in the course of performing cures."[20] The healers that continued to practice were excluded from the Brahmanic social structure:

> These wandering physicians, shunned by the hierarchy of mainstream society, came in contact with groups of heterodox ascetics, or *sramanas*, who were more receptive to the healing arts as well as to a more observational orientation. Both groups were indifferent or even antagonistic towards the orthodox scholastic tradition, and further development of medical knowledge and the healing arts found a receptive home amongst the sects of *sramanas*, which included Buddhists, Jains, and Ajivakas. In due course, the healers became indistinguishable from the other *sramanas*, and the use of empirical procedures

and direct anatomical observational techniques contributed to a vast storehouse of medical knowledge, which supplied the Indian medical tradition with the precepts and practices of what later came to be known as *Ayurveda*.[21]

As in the case of the novelist John Cleland, who was a writer for the Company before he became a renowned author, the British East India Company furnished a place for many an adventurer to flee from domestic troubles and throw themselves on the fortunes of foreign travel and foreign trade. But the Company also attracted amateur scientists who "perceived India to be a vast, unexplored territory" holding out the "promise of totally new flora and fauna" as well as the "possibility of developing their careers as 'scientists.'"[22]

One of these, J.Z. Holwell, discusses in a paper sent to the Royal Society in 1767 the methodology of inoculation in great detail, noting that the:

> Art of Medicine has, in several instances, been greatly indebted to accident; and that some of its most valuable improvements have been received from the hands of ignorance and barbarism: a truth, remarkably exemplified in the practice of inoculation of the small pox. However just *in general* this learned gentleman's remark may be, he will as to his *particular reference*, be surprized to find, that nearly the same salutary method, now so happily pursued in England, (howsoever it has been seemingly blundered upon) has the sanction of remotest antiquity. . . . If the foregoing essay on the Eastern mode of treating small pox, throws any new and beneficial lights upon this cruel and destructive disease or leads to support and confirm the present successful and happy method of inoculation, in such wise as to introduce, into *regular and universal practice, the cool regimen and free admission of air*, (the contrary having proved the bane of millions) I shall, in either case, think the small time and trouble bestowed in putting these facts together most amply recompensed.[23]

Clearly, the transmission and dissemination of this practice predated Montagu and Jenner by a number of centuries, according to these writers, and had very specific genealogies in Vedic scriptures.[24] Holwell elaborates on Coult's reference to Vaidhya[25] practice:

> The sagacity of this conclusion, later times and discoveries has fully verified, at the period in which the *Aughtorrah Bhode* scriptures of the Gentoos were promulgated, (according to the Brahmins three thousand three hundred and fifty six years ago) this disease must have been of some standing, as those

Scriptures institute a form of divine worship, with *Poojahs* or offerings to a female divinity, stiled by the common people *Gootec Ke Tagooran* (the goddess of spots), whose aid and patronage are invoked during the continuance of the small pox season, also in the measles, and every cutaneous eruption that is in the smallest degree epidemical. Due weight being given to this circumstance, the long duration of the disease in Indostan will manifestly appear; and we may add to the sagacious conjecture just quoted, that not only the Arabians, but the Egyptians also, by their early commerce with India through the Red Sea and Gulf of Mocha, most certainly derived originally the small pox (and probably the measles likewise) from that country, where those diseases have reigned from the earliest known times.[26]

Holwell is also careful to note that the Vaidhyan practitioners with whom he consulted were quite aware of an early form of germ theory that constituted the etiology of this disease:

They lay it down as a *principle* that the *immediate* cause of the small pox exists in the mortal part of every human and animal form; that the *mediate* (or secondary) *acting* cause, which stirs up the *first*, and throws into a state of fermentation, is multitudes of *imperceptible animalculae* floating in the atmosphere; that these are the cause of all empideidemical [*sic*] diseases, but more particularly of the small pox.[27]

Complaining that the "usual resource of the Europeans is to fly from the settlements, and retire into the country before the return of the small pox season," Holwell gives an eloquent defense of a medical practice that predates both Turkish or European adoption and expands its etiology to include places that, by the Victorian era, had been dismissed as barbarous, primitive, and, most importantly, sites where the disease originated.[28] The notion that methods of preventing the transmission of smallpox may also have originated from these sites simply did not register with Victorian cultural consciousness.[29]

Early prophylactics to this powerful disease were transmitted by East India Company members to London, but no one read their letters—or, if they did, they seemed to have been dismissed as the fantasies of "common voyage-writers"—until Montagu made it her own noble cause. In a preternaturally inoculative fashion, Indian methods of smallpox inoculation had already been incorporated into the British corpus before it became a Levantine cause celêbre and was officially lodged in the scientific annals of the Royal Society. Paradigms for inoculation existed in the British scientific and medical archive; yet they were ignored by a population of Britons whose xenophobia seemed to increase with every new Indian purchase acquired through the military contests that now characterized East India Company pursuits. The value of a British archive of scientific epistemology may thus have been defined by xenophobia: one could accept Lady Mary Wortley Montagu's letters from Constantinople, but correspondence from India—a place increasingly defined by primitivism by the time of Jenner's experiments—could be safely and justifiably disregarded and forgotten. Edward Jenner, then, did not have to openly acknowledge the existence of a similar practice when he created his successful cowpox vaccine by the end of the eighteenth century. Yet the disease raged in late nineteenth-century Britain, killing as many as 50,000 people in 1871. The question is: why? The "answer" lies in a complicated ideological amalgam of class, race, gender, and imperial identity that, I argue, is driven by xenophobia.

Nadja Durbach argues that the anti-vaccination movement that started in response to the 1853 vaccination acts was a defining phenomenon for late Victorian England. Pointing out the distinction between inoculation and vaccination, she uncovers the ways in which these two practices are ideologically different: "inoculation was generally performed by paramedical personnel such as Nanny Holland, who were in direct competition with vaccinating doctors. Indeed, doctors consistently depicted inoculation as a feminine, foreign, folk practice in contrast to vaccination, which they constructed as masculine, English, and expert."[30]

These beliefs replicate William Wagstaffe's fears of consuming foreign bodies. Jenner's experiments with vaccination in the 1790s paid little attention to the practice of variolation that was common with lay practitioners who used live cultures to inoculate their patients except to highlight the dangers of this procedure. But Jenner paid no attention to the inoculation process defined by East India Company members. J.Z. Holwell had addressed the Royal College of Physicians in London well before Jenner's experiments (1767), describing in minute detail the method used by traveling Vaidhyas for at least the past 150 years (according to his letter), which would fix inoculation in India in the second decade of the seventeenth century, about the time of the first round of East India Company correspondence:

> Previous to the operation the Operator takes a piece of cloth in his hand and with it gives a dry friction upon the part intended for inoculation, for the space of eight to ten minutes, then with a small instrument he wounds, by many slight touches, about the compass of a silver groat, just making the smallest appearance of blood, then opening a linen double rag (which he always keeps in a cloth around his waist) takes from hence 2 small pledgit of cotton that had been emerged with the variolous matter, which he moistens with two or three drops of the *Ganges* water, and applies it to the wound, fixing it on with a slight bandage, and ordering it to remain on for six hours without being moved, then the bandage to be taken off, and the pledgit to remain until it falls off itself. . . . from the time he begins the dry friction, to the tying of the knot of the bandage, he never ceases reciting some portion of the worship appointed, by the *Aughtorrah Bhade*, to be paid to the female divinity before-mentioned. . . . The cotton, which he preserves in a double callico rag, is saturated with matter from the inoculated pustules of the preceding year, for they never inoculate with fresh matter, nor with matter from the disease caught in the natural way, however distinct and mild the species.[31]

I have quoted from Holwell's letter at length to point out that far from using live smallpox cultures collected from infected people, the Vaidhyan method emphasized the danger of such forms of inoculation.[32] It may very well be that the cowpox vaccination that Jenner developed (although there is evidence that cowpox had been used for inoculation before Jenner's experiments) had a higher rate of success than older methods of variolation practiced by both Brahmins and Turks.[33] The point is not which method ultimately prevailed. Rather, I am interested in the ideologies that underpin Jenner's complete lack of interest in any previous body of knowledge about smallpox, especially given his apprenticeship to John Hunter (1770), who was elected as Fellow of the Royal Society in 1767, the year Holwell (who himself was a Fellow of the Royal Society) had published his observations. Returning to Durbach's contention that inoculation was now represented as a "feminine, foreign, folk practice," and drawing on Wagstaffe's earlier contention that inoculation was practiced "by a few *ignorant women*, amongst an illiterate and unthinking People," it would seem as if the distinction between inoculation and vaccination was predicated not as much on empirical evidence as on ideologies of gender, race, and nationalism. Even if Jenner had had access or had acknowledged access to earlier forms of inoculation, he had to render the method English. Hence the historiographical attention paid by Jenner in Berkeley to the milkmaids, who were apparently immune to smallpox according to Jenner's observations. The English milkmaid, Sarah Nelmes, from whose blistered hand he extracted the cowpox "matter," and even the cow Blossom, from whom Sarah Nelmes had contracted the disease and whose hide still hangs on the walls of St. George's Hospital where Jenner trained, rewrote both Indian and Turkish scenes of medical innovation as English.[34] Variolation and inoculation, originally an either Indian or Levantine method, were thus rendered the pathogenesis of British epistemology that transformed those processes into the far more constructive practice of vaccination.[35] Replacing the Vaidhyan prayer (*poojah*) by *Gootec Ke Tagooran*, the goddess of spots, whose patronage was sought at every outbreak of smallpox in Bengal, and the invocations to Krishna (who is famously associated with milkmaids) to whom Vaidhyans would also plead, with the milkmaid lore that Jenner appropriated, Jenner created the conditions in which vaccination could now be read as "masculine, English, and expert." It is, perhaps, a small irony that cowpox is indigenous to England, and that shortly after Jenner published his "discoveries," the demand for cowpox matter grew exponentially so that the proper transportation of the lymph became a powerful issue.[36] Moving from the dangerous urban topography that made it possible for smallpox (and smallpox inoculation) to be imported and interpolated into the body of the metropole, vaccination thus replaced inoculation and relocated the entire method as part of a pastoral landscape that generated its own indigenously produced agent against disease.

Nevertheless, resistance to smallpox inoculation in Victorian England found compelling voices. William Tebb, speaking in 1881 at the Second International Congress of Anti-Vaccinators, appealed primarily to a deeply rooted sense of Englishness to promote his position, claiming, "Compulsory medicine . . . is opposed to the ancient constitution of England, and is, therefore, a gross infraction of the liberty of the Citizen and of parental rights."[37] In 1896 Walter Hadwen, a doctor and an ardent antivaccinator, delivered the following inflammatory speech to a crowd of passionate believers:

> It is not a question merely of the health but of the very lives of the children which are at stake in this matter; and I believe that the present century shall not close until we have placed our foot upon the dragon's neck, and plunged the sword of liberty through its heart. . . . Yes, we are going forward with the "crazy cry" of liberty of conscience upon our unfurled banner, and we never intend to rest until we get it.[38]

The power and weight of these arguments clearly depends on a strong sense of national identity but one that conceives of Englishness as its robust center. The place of the Irish, the Scots, and the Welsh—to say nothing of South Asia—is strangely eccentric to the discourse of antivaccinationists, especially considering that smallpox vaccination was mandatory in colonial India. Despite the rhetoric that Tebb, for example, used to unite his cause to "the Colonies and the whole of Europe," little attention was paid to inhabitants of those colonies.[39] It seems, then, that the threat was most virulent at the domestic center of the metropole, whose susceptibility to infection needed not only state-sanctioned legislation but also a strong xenophobic voice to police its boundaries.

But what were the boundaries of imperial Britain? Anxiety over colonial borders erupted in a number of different discourses, and in particular, the arena of disease and public health was open to an especially xenophobic rhetoric. Alan Bewell identifies how disease defines boundaries:

> Cholera crossed many of the boundaries—cultural, geographical, and climatic—that were thought to exist between Britain and its colonial possessions, and by so doing it challenged those boundaries and led to their reconceptualization. It changed how the British saw themselves and their place in the colonial world. Significantly, this new understanding emerged in tandem with a new conception of India, which was now perceived as the cause, the geographical locus, and the primary exporter of a modern plague.[40]

Although Bewell is referring specifically to cholera, the same arguments can be made for smallpox's environmental footprint and its newly renegotiated geographic locus. Victorian England seemed to have rewritten smallpox as an environmental disease. For example, Alfred Russel Wallace believed that certain types of environmental conditions caused smallpox: "foul air and water, decaying organic matter, overcrowding, and other unwholesome surroundings." Many believed that even if smallpox was transmitted from person to person, under the right conditions it could spontaneously generate.[41] For Wallace and other anti-vaccine supporters, the very idea of integrating disease that was produced by dirty environments as a prophylactic measure was counterintuitive and nonsensical. More important, vaccination signified a dangerously unpatriotic act: a willing renunciation one's English fairness, an enthusiastic fall into alterity. Interestingly, about this time the idea of disease had shifted from having a mercantile mobility—Defoe's representation of the plague traveling on ships laden with goods from the Levant—to being fixed in the landscape, typically ones located in India. Writing for the *Lancet*, James Martin argues

> The cholera epidemics which have ravaged various parts of Hindustan since
> 1817, have always originated in and issued forth from India, but not, to my
> knowledge, been imported into India by ships from infected countries. . . . It
> may therefore be inferred, that the cause of the disease, however latent or
> submerged for a time, is never actually absent from the soil of India, or from
> some of its localities.[42]

Once again the rhetoric connecting the environmental origins of smallpox that anti-vaccinators used to their advantage turns this specific reference to cholera into a paradigm for smallpox as well. As Durbach notes, the distinction between contagious diseases and ones with environmental derivations was often blurred.[43] The boundaries of Britain were continually vexed by the project of its imperial expansion. For Victorian Britons caught up in the anti-vaccination movement it made sense to keep the limits of disease at a convenient distance, if only to foreclose any possible contact with the foreign techne of inoculation no matter how much it had been recast as an English practice. As long as one attended to cleanliness and purity, one could safely suppose to have extirpated the cause of smallpox so that inoculation was unnecessary and superfluous.[44] Yet this was also the moment of Britain's cultural triumph as the civilizing missionary. Figures like St. John Rivers, "a more resolute, indefatigable pioneer never wrought," who, in Charlotte Brontë's representation, "labours for his race" and "hews down like a giant the prejudices of creed and caste that encumber it" loomed large in literary representation.[45] These contradictory acts may be usefully understood as driven by xenophobia, which isn't simply a fear of the foreign. Xenophobia operates as a fetish: as something in which we invest and cathect a good deal of cultural meaning in order to organize and clarify our own culture and nation as something distinct.[46] It was possible, then, for the anti-vaccination movement in Victorian Britain to launch its often vitriolic invective against the Parliamentary body that made smallpox vaccination mandatory because there was already in place, albeit nearly invisible, certainly unread, fragmented epistolary records of the Indian practice of inoculation.

The antipathy toward vaccination articulated at the end of the nineteenth century marks a distinct shift of Britain's imperial aims. Methods of inoculation forced Britons to suspend, however theoretically, the xenophobia that structured cultural, metropolitan, and civic British identity. The introduction of foreign bodies, denatured or not, into the body as a way of warding off disease appeared to Lady Mary Wortley Montagu perfectly sensible, just as it had to Robert Coult, J. Z. Holwell, and Helenus Scott, another doctor working for the British East India Company, partially because these fearless souls were open to receiving new ideas, but also because their travels placed them in contact zones, to borrow Mary Louise Pratt's term, where ecologies of new diseases and biomedical treatments were often patently visible.

Literary representation of smallpox reflected the perceptive shift from eighteenth- to nineteenth-century Britain. John Cleland's masterpiece, *Memoirs of a Woman of Pleasure*, is probably best remembered for its salacious fantasies of Fanny Hill's life. What's less well known about the novel is its investment in disease. Before commencing her tales of bedroom drama, Fanny Hill describes the conditions that prompted her to follow a life of "pleasure," stating the stark circumstances of her familial situation:

> I was now entering my fifteenth year, when the worst of ills befell me in the loss of my tender fond parents, who were both carried off by the smallpox, within a few days of each other; my father dying first, and thereby hastening the death of my mother, so that I was now left an unhappy friendless

Orphan. . . . That cruel distemper which had proved so fatal to them had indeed seized me, but with such mild and favourable symptoms that I was presently out of danger, and what I then did not know the value of, was entirely unmarked.[47]

Cleland's perfunctory remarks about Fanny's girlhood follow a tradition established by eighteenth-century novelists in which the heroine's orphaned circumstances are hastily explained, thus leaving her open, as it were, to a series of adventures that would hardly have been sanctioned by responsible parents. It is no mystery, then, that Fanny's status is a consequence of smallpox, given the ubiquitousness of the disease. Fanny is "seized" by this "cruel distemper" which, far from killing her or permanently imprinting her with the marks of its temporary possession, leaves her "entirely" untouched. This small detail acts as the inoculative moment: Fanny's interpolation of the disease that has had such fatal consequences for her flawed parents—her maimed father, mender of fishing nets, and her beleaguered schoolmistress mother—renders her infinitely more valuable in the marketplace of social and sexual congress. Rather than remaining in Lancashire to replicate her parents' "scanty subsistence," crippled by her father's legacy of castration, Fanny is thus able to seek her fortune elsewhere. Having been infected by Esther's representations of London's fine sights, she decides "all which [I] imagined grew in London, and entered for a great deal into my determination of trying to come in for my share of them." As she is "entirely" untouched by the marks of smallpox, so Fanny is "entirely taken up with the joy of seeing myself mistress of such an immence sum" (eight guineas, seventeen shillings in silver). The "mild and favourable symptoms" that characterize her bout of smallpox also seem to inoculate her against the immoral travails of prostitution; her sexual adventures notwithstanding, Fanny accrues a real fortune, a secure marriage, where, "in the bosom of *virtue*, [she] gathered the only uncorrupt sweets."[48]

John Cleland's personal history attests to his proximity to smallpox and, perhaps, to smallpox inoculation. His younger brother, Henry, had contracted the disease as a child, but had survived the ordeal and, after attending Westminster and Oxford, was established in a colonial position in the West Indies.[49] Cleland himself entered the British East India Company in 1723 shortly after he left school; as his chief biographer, William H. Epstein, somewhat poetically notes:

> That at age seventeen, when he could have been enrolling at Oxford or Cambridge, he was boarding the *Oakham* is the unexplainable mystery of his young life. Born into a family of aggressive social climbers, matriculated in the most prestigious of England's aristocratic Public Schools, Cleland appears to have been cruising in the mainstream of a life style which should never have carried him to the teeming tropical port of Bombay. But something, some inescapable or accidental event . . . diverted the first stage of his journey and channeled him, unprepared for a new tack, into unfamiliar waters.[50]

As if to demonstrate Hal Gladfelder's observation that "Cleland in fact is far more a phantom than his fictional persona Fanny Hill," Epstein's musings about what might have steered Cleland to Bombay replicate Fanny's own accidental journey to the port of London "teeming" with its own brand of diseases.[51] Bombay Island, according to Epstein's projections, was "haphazard chunks of land, debris cast aside or unassimilated by the mainland," saturated with death and disease, a place " 'from which exhalations from the putrid fish or *koot*, with which the lands were manured,' were thought responsible for the epidemic diseases which periodically ravaged the population."[52] These epidemics, Epstein notes, Cleland was fortunate to have evaded, considering that the climate "claimed the lives of well over half the Company's servants."[53] In a much more detailed study of climate and disease, Robert Markley notes that the "British outpost at Bombay . . . was marked by such high rates of mortality for merchants and sailors that it forced the British to reassess some of their fundamental assumptions and values about the relationships among climate, ecology, and human health."[54] Epstein's musings, while not paying particular attention to the climatological or ecological reasons for the high mortality rate that Markley rightly argues, are nevertheless evocative of the xenophobia and gynophobia characterizing public health.

Baffled by the mystery of what could have compelled Cleland, armed "with a classical education and a father who could afford to place him closer to home," to "fight for the sovereignty of this little island," Epstein speculates that perhaps it was the mother, whose family had lived and died in India

that drew the young man thither. Of the fact that Cleland's sister Charlotte joined him and survived as well as he, not to mention the example of
his brother Henry, Epstein is curiously silent, but reading Epstein's un-
flattering account of Cleland's mother, there seems to be some general
implication that impropriety somehow breeds resistance to disease. At
any rate, Gladfelder notes that Cleland's friendship with fellow East India
Company servant Michael Carmichael included a shared reading of the
titillating 1655 *L'École des filles* and an early collaboration on *Memoirs of
a Woman of Pleasure*. Cleland successfully navigated the dense hierarchy
of the East India Company as he advanced from common foot soldier to
gunners' assistant to attorney to writer, a position "which not only gave
him a sure foothold on the bureaucratic ladder but allowed him to engage
privately in trade on his own behalf." Cleland's tenure resembles Elihu
Yale's in that he used his position to defend Hindu merchants against the
Company's machinations, to shelter a slave woman in his house, and to
entertain possibilities of writing his own "stark naked truth" of a harlot's
progress.[55]

John Cleland spent the period between 1728 and 1740 working for the
company in Bombay, years where he may very well have known about and
observed the practice, given the prevalence of both smallpox and inoculation before the introduction of vaccination in the nineteenth century.[56] He
returned to England in 1741, was arrested in 1748 and put into the Fleet
Prison for debtors where, at least according to most histories, he wrote
Memoirs of a Woman of Pleasure, which was published later that year.[57]
Gladfelder, however, raises the intriguing specter of the "Bombay Fanny
Hill," the colonial original of the English novel: a collaborative effort
spurred by Michael Carmichael's challenge to write the life of a prostitute
with no coarse words, spiced with their forbidden pleasures in reading *L'École
des filles* and indulging in other illicit amusements with native slaves. Cleland's return to London was not triumphant. Like a soberer version of Fanny
Hill, Cleland wandered, and the years between his return and his arrest
"were years that marked a radical change in his life's direction, and the be-
trayals and frustrations that infuse and perhaps disfigure his later work all
lead back to this period when Cleland—in the words Samuel Johnson wrote
of his scapegrace friend Richard Savage—'having no Profession, became by

Necessity, an Author.'"[58] London's Smallpox Hospital was built in 1746 with the exclusive purpose of inoculating and treating the poor, and remained the only place where the indigent population could receive free inoculations. Surely Cleland, fresh from his tenure in Bombay, having been refused any part of the family fortune and languishing in prison over a £840 debt, would pay attention to the horrors of a dangerous and defacing disease facing a young orphaned girl and would include this inoculative fantasy as part of his erotic imaginary. Like Fanny, Cleland survives diseases rampant in Bombay, but unlike Fanny, whose unmarked face assures her of a proper domestic future, Cleland's first unmarked script "Fanny"—the manuscript of the *Woman of Pleasure* he had produced with Carmichael— was gone: "a conjectural urtext whose relation to the published text is unknowable," according to Gladfelder, but one that produces the disfigured face of the published *Memoirs of a Woman of Pleasure*.[59]

Literary representation of inoculation makes a definitive turn in the nineteenth century. One of the grand hysterical novels of the fin de siècle, Bram Stoker's *Dracula*, reflects much of the xenophobic resistance to vaccination. Although *Dracula* doesn't offer metaphorical allusions to vaccination, as Durbach contends, it is a novel that is concerned with inoculation and its various threats. The novel's apprehension of the generalized Victorian anxiety over women's "concealed inner recesses" harboring polluted blood reaches a frenzied climax in the killing of the vampiric Lucy Westenra.[60] Although much of the scene enacts the legitimate consummation of Lucy's "marriage" to Arthur Holmwood, a ceremony and ritual that reinstates his proper place as the Victorian husband and restores sexual civility to Lucy, her overdetermined healing speaks to a curiously conflated set of anxieties over borders and boundaries, the environmental etiology of disease, and of the dangers of inoculation. Bewell argues that the definition of contact Britain had with India was problematic: "Spread along the main transportation and commercial arteries of the nineteenth century—by river, sea, road, and later by railway—cholera mapped the many lines of communication between Britain and its colonial possessions."[61]

Cholera replicates the routes taken by smallpox in Defoe's *Journal of the Plague Year*; the problem with colonial and commercial contact is that diseases that were represented as originating from India were now capable of moving into the corpus of the metropole. In much the same way, Jonathan Harker's promising voyage to the Castle Dracula in Transylvania results in the proliferation of vampiric disease that travels by "river, sea, road, and . . . railway" to the very heart of modernity.

Jonathan Harker battles the specter of disease with an arsenal of tools that are "nineteenth-century up-to-date with a vengeance": his shorthand, his Kodak, his ordnance survey maps, and the institutions where he seeks information (like the British Museum) are all used to confine Dracula's uncanny powers within the walls of his castle where they can remain relatively ineffective, incapable of spreading westward. But they fail. The British Museum has no maps of Transylvania or the Castle Dracula, and certainly not any maps that compare with the ones produced by the military division of the Board of Ordnance that was first commissioned to charter the Scottish Highlands in order to control Jacobite supporters.[62] The dearth of information available at the British Museum seems especially trenchant compared with the wealth of British knowledge the Count commands:

> In the library I found, to my great delight, a vast number of English books, whole shelves full of them, and bound volumes of magazines and newspapers. A table in the centre was littered with English magazines and newspapers, though none of them of very recent date. The books were of the most varied kind—history, geography, politics, political economy, botany, geology, law—all relating to England and English life and customs and manners. There were even such books of reference as the London Directory, the "Red" and "Blue" books, Whitaker's Almanac, the Army and Navy Lists, and . . . the Law List.[63]

The sheer variety of texts housed in the Count's library embarrasses Jonathan Harker's position, and he is reduced to showing photographs of the Count's new property in Carfax while the Count reads a Bradshaw, familiarizing himself with railway timetables.[64] Thus Jonathan's tools facilitate Dracula's invasion of London. Even Jonathan's shorthand fails as a techne to counter the nightmares of the past that threaten to kill "mere modernity." Dracula reads through Jonathan: he intercepts Jonathan's calls for help and replaces them with his own cryptic missives. Stripped quite literally of any suit he has—his clothes, his agency—Jonathan, unmanned, is left alone in the castle.

Only after Dracula has inoculated London—after he has planted himself in Carfax and opened Lucy Westenra and Mina Harker's veins—do these tools begin to recover their use. The trains and steamships and railway carriages swiftly pursue Dracula back to his Transylvanian lair. Two knives effectively exterminate Dracula: Jonathan Harker's *kukri* and Quincey Morris's Bowie. The latter knife with its characteristic cross, wielded by the valiant Texan, obviously alludes to the promising technologies produced by new Christian "civilized" frontiers. Jonathan Harker's *kukri* is a little more complicated. Curved like a scimitar, *kukris* were favored by the Nepali Gurkhas in their series of conflicts with the British East India Company that eventually culminated in a war earlier in the century (1814–1816), which is when these knives made their way into England. *Kukris* have a historical connection with the famed eleventh-century Damascus swords fashioned of *wootz*, an early form of carbon steel that was manufactured in India and exported to Damascus from the third to the seventeenth century.[65] Brandishing a weapon that was far more likely to be found in Dracula's armory than in any English store, Jonathan's triumphant return to the scene of his unmanning is made possible by this inoculative appropriation of older Indian technologies.

Shortly before his death, Quincey Morris points to Mina Harker, exclaiming "Now God be thanked that all has not been in vain! See! the snow is not more stainless than her forehead! The curse has passed away!"[66] In an ill-advised moment, Van Helsing, posing as the brave band's erstwhile priest, presses a "piece of Sacred Wafer" on her flesh, which far from protecting her from further visits by Dracula, indelibly marks her face: "it had seared it—had burned into the flesh, as though it had been a piece of white-hot metal." Understanding her position, Mina cries "Unclean! Unclean! Even the Almighty shuns my polluted flesh! I must bear this mark of shame upon my forehead until the Judgment Day" while Van Helsing tries to comfort her "And oh, Madam Mina, my dear, my dear, may we who love you be there to see, when that red scar, the sign of God's knowledge of what has been, shall pass away and leave your forehead as pure as the heart we know."[67] Mina has been "polluted" by Dracula. This scene is most commonly read as the illicit sexual exchange of fluids between Mina and Dracula that finalizes Jonathan's cuckolding. Van Helsing is not a priest; his layman's attempt to wield the host as a means of protecting Mina fails spectacularly as he ends up branding her with the wafer. As a doctor, however, he succeeds in marking Mina as the inoculated subject, the body that has ingested the horrifying disease and become the unclean other. It is important to recognize that Mina is not simply contaminated but is inoculated by Dracula's disease. Unlike Lucy, what Mina has introjected in her exchange with Dracula is the ability to harness Dracula's disease against himself. Together with her adamant resolution to retain her own agency, she exploits vampiric knowledge to contest Dracula's power.

Until Van Helsing scars Mina with the communion wafer, making her visibly recognizable as part of Dracula's ilk, he is almost cruel to her and Jonathan. He answers Jonathan's concerns about Dracula's imminent arrival with a brutal reference to the night before: "'Do you forget,' he said with actually a smile, 'that last night he banqueted heartily, and will sleep late?'"[68] Mina's red scar—an outward sign of "God's knowledge"—echoes an earlier moment when Mina transforms Jonathan's journal into an "outward and visible sign" of their marital trust.[69] These references to Catholic catechism, particularly strange given Jonathan's status as an "English Churchman," insist on god's grace as the final judge of bodily purity, even if the men are not above exploiting Mina's inoculated body in order to hunt and kill Dracula, the source of social disease. Mina's restoration to god's grace and Jonathan's act of ritual killing reflect the ways in which Victorian ideologies of female sexuality were more about assimilating men to the new conditions of the labor market, as Sally Shuttleworth argues.[70] *Dracula* concludes with a scene that privileges domesticity over the avant-garde forays into unfamiliar territory. Jonathan's manhood is secured not by his inheritance of Hawkins' fortune but by the fact that Mina has forsaken her somewhat dubious ambition to venture into the world of journalism for the infinitely more proper joys of motherhood. The fact that not a single document exists to authenticate their experience is of no consequence; rather, the importance shifts to the "bundle of names" that "links all our little band of men together," Quincey Harker.

Although the advertisement in *The Woman at Home* was primarily for Watkins product and not necessarily for treatments of smallpox, such advice did exist in medical journals. The *Cyclopaedia of Practical Medicine*, for example, recommends taking a solution of cream of tartar in "sufficient doses" and also mentions that this should be given to patients who suffer from the disease regardless of its etiology (that is, whether it results from inoculation or not).[71] Cream of tartar was an effective emetic, but it may be worth considering its other properties in order to address the politics of its deployment as a valuable cure for smallpox. Allegedly first isolated by the Persian alchemist, Jabir Ibn Hayyan circa 800 AD, the tartaric acid encrusting barrels of wine was later folded into modern chemistry in 1769 when the Swedish chemist, Karl Wilhelm Scheele, collected the tartaric acid flakes, neutralized them with potassium hydroxide, and developed the substance of kitchen alchemy popularly named cream of tartar.[72] Cream of tartar's interpretive possibilities sharpen when one considers that it's a very efficient stabilizer.[73] As a domestic remedy, emanating from the familiar and familial confines of the home, cream of tartar mixed in water represented the stabilizing qualities of the patriarchal household ministered by the maternal angel in the house. Medical practices like inoculation and vaccination subjected the safety and sanctity of the family to dangerous foreign bodies that had the potential to radically alter the contours of English domesticity.

Anxieties over those contours extended to reflecting on the borders of the national body, and similar anxieties inflected and infected those reflections. Victorian Britain, mindful of its imperial trajectory, seemed to equate the medical practice of inoculation with a cultural one. Thus far from entertaining the introduction of foreign techne into British practice, as Montagu, Coult, Holwell, and Scott were willing to do, many Britons read inoculation as an unpatriotic act, a treasonous introjection of the elements of disease into what they perceived as the healthy corpus of the metropole. Clearly, as a practice inoculation becomes increasingly vexed as Britain professes to turn commercial profit from conquest and colonization into the social missionary work that characterized Victorian cultural imperialism, domestic and otherwise. If, as Bewell has argued, the "global expansion of human travel also made possible the *globalization of disease*," then Victorian xenophobia becomes a strategic defense against infection. Bewell notes that "new colonial disease ecologies" lacked a "uniformity of experiences" and therefore their representation was vexed by the absence of stabilizing structure.[74] Domestic Victorians, fearing the palpable consequences of smallpox epidemics, were prone to locate its etiology elsewhere, thus insuring their cultural immunity to the pathogen.[75] Incorporating pathogenic otherness amounted to declaring oneself the origin of disease, cultural and otherwise, and so advertised cures for smallpox were as mild and putatively stabilizing as cream of tartar dissolved in water.

FIVE

"Plaisters," Paper, and the Labor of Letters

Take of isinglass, half an ounce; Turlington's (or Friar's) balsam, a
drachm; melt the isinglass in an ounce of water, and boil the solution
till a great part of the water is consumed; then add gradually to it,
the balsam, stirring them well together. After the mixture has
continued a short time on the fire, take the vessel off, and spread the
extended silk with it, while it is yet fluid with heat, using a brush for
spreading it.

—*The New Family Receipt Book* (1810)

From the bark are made all kinds of rope, packing cloths, nets, &tc.
And from these, when old, most of the paper, in this country, is
prepared; for these purposes, the fresh plant is steeped four days in
water, afterwards dried, and treated as the cannabis for hemp, to which
it is so similar when prepared, that Europeans generally suppose it to
be the produce of the same plant.

—LIEUTENANT COLONEL IRONSIDE, *Of the Culture and Uses of the
Son or Sun-plant of Hindostan, with an Account of the manner of
manufacturing the Hindostan Paper* (1773)

Jane Austen's novels regarded as a whole: no hidden meanings or
philosophy in them; she only made the familiar and commonplace
interesting and amusing; their style the same throughout; the plots
generally well-sustained, though unsensational; the heroines more
interesting and better drawn than the heroes, but the secondary
characters the best; Jane Austen's narrow range of observation
caused partial recurrences of characters and incidents, but Lord
Macaulay claims that each character is distinct; though her subjects
were commonplace and trivial, her genius has made them bright
forever.

—GOLDWIN SMITH, *The Life of Jane Austen* (1890)

Court plasters appeared in eighteenth-century English culture as a cosmetic device that either covered scars or created them. Strategically placed on the face to emphasize its beauty, they were also deployed to draw attention away from the marks of disease, most commonly smallpox. They were often called "patches" and were used by women of the court with great enthusiasm. As a mark of aristocratic fashion, they were also quite expensive. But by the nineteenth century, court plasters—or plaisters—had devolved from decorative patches into a form of bandage, available at apothecaries everywhere, carried about in the pockets of all women, aristocratic, genteel, or bourgeois, and, as the recipe from *The New Family Receipt Book* demonstrates, often made at home.[1] Although the court plasters of both the eighteenth century and the nineteenth century were virtually identical in composition—a piece of silk or taffeta or cotton or even leather was coated on the one side with a mixture of isinglass and glycerin and cut to size or shape—the difference in price and value was immense. Part of this difference had to do with the manufacture of isinglass. Made exclusively from the swim bladders of the Beluga sturgeon, Russian isinglass as it was also called was expensive to import, and the beauty patches that were made from this substance reflected that expense in their price. By the end of the eighteenth century, however, the composition of isinglass changed. Fueled, perhaps, by William Pitt's claim that cod was "British gold," the Scottish engineer and inventor, William Murdoch, discovered a way to process the swim bladders of cod—a fish readily available in British markets primarily because the conditions of the Treaty of Paris that concluded the Seven Years War gave the British exclusive rights to Newfoundland fisheries. This substitute for the more costly sturgeon made isinglass a much more affordable material, substantiating Pitt's claim.[2] The expensive fashion—one that had a practical use, to be sure, but whose real value was in its superfluous application—thus robbed of its exclusivity, became the common property of all and sundry. Because the primary value of court plasters shifted from the aesthetic to the entirely functional, it is, perhaps, no accident that its appliance migrated from the face to the body. Eventually, the vestigial "court" disappeared from the vernacular and "plaister" became the word of common use.

The methods for making common writing paper used for correspondence was remarkably similar to the process Lieutenant Colonel Ironside describes in his 1773 essay submitted to the Royal Society. The "Hindostan" paper-maker uses

> old ropes, cloths, and nets, made from the sun plant, and cuts them into small pieces, macerates them in water, for a few days, generally five, washes them in the river in a basket, and throws them into a jar lodged in the ground; the water is strongly impregnated with a lixivium of sedgi mutti six parts and quick lime seven parts. After remaining in this state for eight or ten days, they are again washed, and while wet, broken into fibres, by the stamping lever, and then exposed to the sun, upon a clean terrass, built for this purpose; after which they are again steeped, in a fresh lixivium, as before.

The pulp is then spread onto screens, formed into sheets, dried, spread on a mat and rubbed with "a piece of blanket dipped in this rice paste water," dried again, and polished with a "globular piece of moonstone granite." Ironside speculated on the scarcity of "substances producing cloths, ropes, and paper," and recommended that the sun plant "be cultivated with advantage in some of the British West India settlements and in other countries destitute of hemp and flax." He also hypothesized that it would not be "improbable, that it may be raised in the warmer climates of Europe, as it ripens here in winter."[3] Ironside's 1773 paper anticipated the hemp shortage in England of the early nineteenth century.[4] Although quantities of hemp were cultivated domestically, there wasn't enough to supply a British Navy constantly engaged in the Napoleonic wars from 1803 to 1815. Much of the hemp was imported from Russia, and Napoleon's 1812 invasion of that country seriously threatened this source. Hemp was important as the principal fiber in making ropes and nets, but it was also used to make paper (though not the same hemp as the sun-plant Ironside describes).

Plasters are unmistakably material, but the process of making them—the various experiments, for example, that Murdoch made to identify another form of isinglass, his training in chemistry and engineering—tends to be forgotten. Likewise the labor that produces high-quality paper is erased from its use: writing letters, poems, novels, plays, recipes, scientific treatises, and the like, our attention is drawn to what is signified on the material rather than the material itself. In this chapter I focus on the ways human intellect is commodified: that is, in Marx's definition, represented both as a material thing and as a mystification, a thing that is embedded in social relationships. This chapter understands "analogy" as a focus on language's relation to the world; that is, analogy entails a material "fact," such as plaster or paper, and a discursive or social "event"—the complicated status of labor as commodity, as the "thing" that is entrenched in social relations and social circulation. Claudia Johnson's advice to pay attention to "what has always been before us" is germane to reading the connection between material artifacts and human intellect but with a caveat: I focus on the things that are hidden in plain sight.[5]

Plasters and Paper

Probably the most famous literary use of court plasters is found in Jane Austen's novel, *Emma*, in a particularly poignant scene where Emma's poor judgment of character and class alignments is most visibly and materially brought to her attention. Long after Mr. Elton has married and, indeed, some length of time after he has publicly humiliated her, Harriet Smith pays Emma a confessional visit, bringing with her a package.

> She held the parcel towards her, and Emma read the words *Most precious treasures* on the top. Her curiosity was greatly excited. Harriet unfolded the parcel, and she looked on with impatience. Within an abundance of silver paper was a pretty little Tunbridge-ware box, which Harriet opened: it was well-lined with the softest cotton; but, excepting the cotton, Emma saw only a small piece of court plaister.[6]

Buried under a wealth of wrapping is this unprepossessing article of which Emma at first has no recollection.[7] Prompted by Harriet, however, Emma, now penitent, remembers "cutting the finger, and my recommending court plaister, and saying I had none about me . . . and I had plenty all the while in my pocket!" (367). Both Harriet Smith, the illegitimate daughter of a tradesman, and Emma Woodhouse, the privileged daughter of landed gentry, keep lengths of court plaster in their pockets, but it is only Harriet who makes visible use of the tape. Delighted at first that she can accommodate Mr. Elton's needs with this pedestrian piece of plaster, she cuts a piece "a great deal too large" so that Mr. Elton has to cut it smaller; her eagerness to please Mr. Elton, embodied by the inappropriately large length of plaster, has to be disciplined and whittled down to size, the remainder of which he toys with before returning it to Harriet. She then raises the stakes of its signification and revalues the abject and discarded strip as one of her "most precious treasures." As such, this residual "relick" of her feelings for Mr. Elton must be utterly destroyed even when Emma, who has "secretly added to herself 'Lord bless me! When should I ever have thought of putting by in cotton a piece of court plaister that Frank Churchill had been pulling about!—I was never equal to this'" and mindful of restoring Harriet back to solid, practical ground, asks Harriet to keep it, reasoning that it "might be useful" (367–368). But Harriet is unable to extricate the court plaster from its wallow of sentiment and restore it to its proper place as a useful item, and declares she should be happier without it for it "it has a disagreeable look" to her. (368). It is this "look" that situates Harriet squarely in the material sphere. Just as the value of court plasters drifted from the confines of the court to the counters of trade culture, and just as their use migrated from the face to the body, so Harriet, as the material body, embodies the passage of court plaster from the face to the body, from beauty to health, from the symbolic to the material. Her face, with "a fine bloom, blue eyes, light hair, regular features, and a look of great sweetness," can't be "wasted on the inferior society of Highbury and its connections," or so Emma thinks, fancying a future of refinement for her friend (22–23). But her fate is ruthlessly determined. Distracted by these relics that are now neither symbols nor things but nevertheless preserve their "looks," she inexorably confirms the ways in which her socially illegitimate body confines her to the realm of the mere material.

The other item resting in its bed of cotton in the Tunbridge-ware box "is something still more valuable . . . because this is what did really once belong to him": the end of an "old pencil—the part without any lead," discarded by Mr. Elton because it is perfectly useless to "make a memorandum . . . about spruce beer" and therefore "left upon the table as good for nothing" (367–368). What is good for "nothing" for Emma and Mr. Elton—Emma has "not a word to say for the bit of old pencil" while Mr. Elton leaves it on the table without a further thought—is clearly valued as something by the hapless Harriet, even more than the remnant of court plaster that still, after all, retains its use. This leadless stump, snatched as soon as she "dared . . . and never parted with again from that moment," accompanies the plaster into the fire, and "there is an end, thank Heaven! of Mr. Elton" (368–369).

The exchange between Harriet and Emma articulates a moment of *verba volant scripta manent*. For Emma, these scenes have long since disappeared into their semantic past; for Harriet, trammeled by her sentiments for Mr. Elton that are kept alive even though she knows "it was very wrong of [her] . . . to keep any remembrances, after he was married," they are always terribly present. What Harriet keeps packaged, however, is no clear script of a love—no matter how unrequited—now lost. There are no letters, no form of writing that documents past moments whether or not they have to do with courtship or love, which is odd given the preponderance of letters and letter writing in the novel. Rather, what she preserves are the residues of *scripta*: the court plaster, metonymically linked to writing because it is the remainder of the tape with which Mr. Elton has used to bind the wound made by Emma's new penknife, the end of a pencil whose lead he cut away until "it would not do, so [Emma] lent him another" (368). Hopelessly bound to the materiality of what these relics signify and not their semantics, Harriet's social position remains equally mired in abject material, no matter what grand plans Emma may fabricate for her friend. Even if she has her flights into the world of abstract representation as when she transforms plaster and pencil to icons of love, she is disturbed by their "look," and so her icarian return to the mundane situates her undeniably as the natural daughter of trade. Her participation in the realm of courtship is similarly literal: she treasures a plaster stripped of its aristocratic association and a pencil emptied of its function. Emma, by contrast, has access to the materials of proper writing—penknives and pencils—that represent an entire abstract world of writing, but, appropriately, she keeps the prosaic plaster hidden in her pocket.[8]

The issues of hands, labor, letters, and paper have come together earlier in the novel, when Emma first introduces Harriet to the possibility of expanding the horizon of her marital expectations. Emma devises a course of self-improvement for Harriet but lacks the determination to get beyond the "few first chapters." Left rudderless, Harriet's intellectual activity is limited to "the collecting and transcribing all the riddles of every sort that she could meet with, into a thin quarto of hot-pressed paper" (74). The material defines Harriet's first exposure to the world of ideas: having been encouraged by her head teacher, Miss Nash, to assemble such a collection, Harriet's "very pretty hand" ensures that it "was likely to be an arrangement of the first order, in form as well as quantity" (74). The substance of the book is in its material dimensions—the embellishments of "cyphers and trophies," the handwriting, the copied charades, the number of charades, even the type of paper are all chosen for their appearance. Hot pressing was a process that produced a smooth, highly polished, glossy paper with little surface texture. It was ideal for watercolors and other forms of pencil art, but not as useful for writing because the surface provided little absorbency for ink. Harriet herself is incapable of any absorbing interpretation: she values the charades not for their meaning but for their quantity. The first charade she receives needs to be explicated: its riddle, courtship, remains unreadable to her until Emma makes its meaning clear and declares that Harriet herself is its "object," adding that this connection to Mr. Elton will give her every "thing that [she] want[s]—consideration, independence, a proper home" and, most important, will "confirm [their] intimacy for ever" (78–79). Emma's fantasies for Harriet, impelled by her misreading of the "object" of Elton's charade, are intended to liberate Harriet from the limitations of illegitimacy. The only natural confirmation available to Harriet, however, is her marginal relation to the court of the landed gentry. She can collect and copy charades but cannot read them; her relation to writing is restricted to the same material dimensions as the ornamental paper she uses for her quarto.[9]

Harriet's abjection is curious given the amount of money she invests both in this quarto and in the parcel containing her "most precious treasures." Wrapped in "an abundance of silver paper," the Tunbridge-ware box is "well-lined with the softest cotton," something Emma notes is too valuable to waste on a mere piece of court plaster. While the price of isinglass may have contributed to the humble position court plasters now occupied in the early nineteenth century, the price of cotton and paper and Tunbridge-ware was still quite high. James Whatman developed wove and hot-pressed paper in the 1750s. This process used a different, much finer wire screen for the paper-making mold (which was called wire cloth) that allowed for an even formation of the pulp fibers without any textural impressions left on the surface of the newly formed paper. The resulting paper was then repeatedly polished. It was the finest paper made in Britain in the eighteenth century, and when Whatman died in 1759, his son (also James Whatman) inherited the Turkey Mill in Kent that produced the paper, which, because of the price of cotton, continued to be quite steeply priced well into the nineteenth century. The prodigal use Harriet makes of costly paper and packaging thus puts her on par with Emma in terms of her capacity to spend.

There were, however, other ways of making paper that cost a good deal less than the hot-pressed wove Harriet uses for her collection, as Ironside explains. The process of making paper from hemp started in England in 1494 (the Magna Carta was printed on hemp paper), and by the end of the eighteenth century, cheap paper made from hemp was popular for correspondence. Austen's novel, written a scant year after the Battle of Waterloo forced Napoleon to abdicate his imperial position and thus end over a decade of war,[10] seems as far removed from the political conditions of warfare, trade, and empire as Goldwin Smith's somewhat patronizing analysis of her writing suggests.[11] Obviously, many scholars have since debunked his Victorian evaluation of her work. The notion that her novels have "no hidden meanings or philosophy in them" is quite ludicrous to modern readers; indeed, the idea of a "hidden" meaning itself is almost farcical to academics that spend their careers disabusing their students of this phrase. In fact, hiding was an accomplishment Austen putatively mastered in her personal life—or so readers like Virginia Woolf wanted to think, and even the fictions spun by her own family elaborate on the myth of the "proper lady," to borrow Mary Poovey's term. According to these reasons, one might therefore believe that many things might be buried, embedded—even "hidden"—in the pages of her novels including the relation between *Emma* and hemp.[12]

Commenting on the paternalistic and patronizing treatment of female writers in her essay, *A Room of One's Own*, Woolf makes a great deal of Austen's skill at hiding. She describes the conditions of Austen's writing as follows:

If a woman wrote, she would have to write in the common sitting-room. And, as Miss Nightingale was so vehemently to complain,—"Women never have an half hour . . . that they can call their own"—she was always interrupted. Still it would be easier to write prose and fiction there than to write poetry or a play. Less concentration is required. Jane Austen wrote like that to the end of her days. "How she was able to effect all this," her nephew writes in his Memoir, "is surprising, for she had no separate study to repair to, and most of the work must have careful that her occupation should not be suspected by servants or visitors or any persons beyond her own family party." Jane Austen hid her manuscripts or covered them with a piece of blotting paper. Then, again, all the literary training that a woman had in the early nineteenth century was training in the observation of character, in the analysis of emotion. Her sensibility had been educated for centuries by the influences of the common sitting room. . . . Yet she was glad that a hinge creaked, so that she might hide her manuscript before any one came in.[13]

Later, Sandra Gilbert and Susan Gubar use the same quote to illustrate their trope for the woman writer, and Jane Austen in particular, and the nineteenth-century literary imagination: the cover story.[14] This form of coverture—the blotting paper with which Austen supposedly covers her manuscripts—was typically made from the same rag content as writing paper, differing only in the fact that it was unsized to ensure its capacity to absorb excess ink.[15] If we disengage the trope's semantic meaning from the material use to which it was put, perhaps a very different understanding of writing emerges. Both writing and blotting paper made from hemp would have been more widely available than cotton or linen paper, especially to young, female, unmarried writers like Austen who were beholden to their fathers for "the spaces they provide."[16] Clearly Austen must have used blotting paper like any other writer; using blotting paper to hide her writing—if that's what she did—provides a compelling prospect of the fate of bourgeois female authors whose talents are absorbed or hidden by the grim reality of their material conditions.[17] Even if Austen, as Betty Schellenberg has so eloquently argued, was not the modest, retiring author divested of any interest in her publishing career, nevertheless future generations of critics—including ones who had a vested interest in recovering her for a feminist posterity—are peculiarly caught up in the image of the proper lady.

The fact that hemp was most widely used for making rope—as in the case of the British Navy—and secondarily processed as paper (that is, in the form of used rope, nets, and clothes) suggests even more figurative possibilities for the place of unpropertied women. Ropes are clear metaphors of bondage—they hold things fast—but they also connect. The sails, sheets, and halyards of East Indiamen were made of hemp as were the ropes used in other forms of their square rigging, an image associated with the promise of goods and luxuries from afar (as in Thomas Rowlandson's 1792 cartoon "The Contrast"). Representations of the fore-and-aft rigging of schooners and sloops suggested the kind of speed and maneuverability also necessary to make foreign connections, even if only in combat. Both in trade and warfare, then, ropes were unmistakable tropes of imperial connection and expansion. Ropes were metonymically associated with slavery as well (although iron shackles were most commonly used to bind slaves) and images of bondage dominated the discourse of abolition in the first fifteen years of the nineteenth century. Embedded—hidden—in the materiality of paper were the material and semantic meanings of "rope"; as the ties that bind or the ties that connect, that oppress or secure, either sense saturating Austen's novel and the labor of Austen's writerly survival.[18] Ironside's search for alternative sources of hemp, tying him to labor in India, also binds him to the representations of female labor, written on the very material of their signification.

The multiple meanings of bondage complicate the novel from its inception. *Emma* opens with an elegy of sorts to the late Miss Taylor, now Mrs. Weston. This melancholy reverie is primarily characterized by the troubling problems the paid companion poses to the eponymous heroine. The "sixteen years" in which Miss Taylor has held "the nominal office of governess"—even if she has been "less as a governess than a friend"—are reviewed by Emma as a "black morning's work":

> She recalled her past kindness—the kindness and affection of sixteen years— how devoted all her powers to attach and amuse her in health—and how she nursed her through the various illnesses of childhood. A large debt of gratitude was owing here; but the intercourse of the last seven years, the equal footing and perfect unreserve which had soon followed Isabella's marriage on their being left to each other, was yet a dearer, tenderer recollection. (4)

If the recollection of the last seven years is "dearer" to Emma, it is because she can expunge the debt she has accrued with a fancy of "equal" footing. Unsaid and unaccounted for are the many hours of labor of the preceding nine years in which Miss Taylor—like Miss Smith—burdened with the name of a trade, has attended not only to Emma's own childhood caprices but presumably to her sister Isabella's as well.[19] Miss Taylor's complete devotion to Emma's needs in sickness and in health recalls less the English governess and more the Indian *ayah*: the abject but indispensable nursemaid/lady's maid of the nineteenth-century Anglo-Indian household.[20] The fact that she has managed to net a husband of "easy competence" when she herself is "portionless" is itself a small miracle given her straitened circumstances, but the fact remains that she has now exchanged her "dependence" on the Woodhouse family for the relative "independence"—in Mr. Knightley's words—of the Weston, has replaced her bonds of servitude as a governess with the more proper ones of a wife.

Because of this, everyday life for Emma has changed: the "difference between a Mrs. Weston only a half a mile from them, and a Miss Taylor in the house" attests to Emma's fondness for her old governess, but, more important, it alludes to the difference in claims to property. Neither Emma nor her father can possibly imagine for Miss Taylor the benefits of having her own home when "this [one] is three times as large," but then neither of them can materialize the liminal place she has occupied in their home for sixteen years to make such a comparison possible. Her labor is invisibly tendered, which enables Emma's "work" to transubstantiate the "debt" (one wonders if she ever actually gets paid) into the more pleasing fantasy of the "perfect unreserve" of their relationship feasible. The geographical difference that half-mile makes perhaps accounts for part of the unpleasantness Emma experiences when she "venture[s] once alone to Randalls": she is conscious for the first time of just how far beyond her reach a simple half-mile has placed Mrs. Weston. It is shortly after this lonely walk that Emma employs another companion in "a Harriet Smith" whom she can "summon at any time to take a walk" and who is "exactly the something which her home required" (25).

Like so many of her novels, *Emma* represents Austen's commerce in the everyday business of writing. Far from being confined to the "narrow range of her observation," as Goldwin Smith claims, *Emma* dramatizes the not-so-everyday issues of intellectual property, particularly as it pertained to women, to the underclass, and to other subjects of the far-flung empire. The expression "every day life" conjures the specter of one of the most renowned social historians and his tour-de-force, Fernand Braudel's *The Structures of Everyday Life*. Braudel intriguingly describes these structures as the "inventory of the possible" whose limits are set by "that 'other half' of production which refuses to enter fully into the movement of exchange." Such a list, drawn from the twin registers of the material and the economic that account for "material civilization," thus helps us to "strip ourselves in imagination of all the surroundings of our own lives if we are to swim against the current of time and look for the rules which for so long locked the world into a stability."[21] Thus the price of plasters and paper occupy a crucial place in the inventory of Austen's world and operate not only as tropes but also as the material effects that mark out the limits of the early nineteenth-century female writer. What were the political conditions that both kept Austen bound to the insular community of rural England and connected her to the remote reaches of the British empire?[22]

You-Me Park and Rajeswari Sunder Rajan's edited collection, *The Postcolonial Jane Austen* addresses many ways of "remapping" Austen in the postcolonial world. The fine essays that constitute this collection are mostly concerned with social relations underscoring various interpretations of Austen's work. My concern is with the material connections *Emma* has to India. James Thompson and William Galperin have offered other glimpses of the double registers of economics and material culture, but with a specific focus on the aesthetic function of the novel in general for Thompson and Austen's novels in particular for Galperin. Thompson argues persuasively for the relation between the novel and political economy, uncovering the ways in which these putatively separate spheres are mutually constitutive.[23] Galperin's reading of *Emma* focuses on the legacy of epistolarity that uncovers the value of competent reading. Galperin invokes another French thinker of everyday life, Michel de Certeau, to read the possibilities of oppositional practices of everyday life that both elude and affix discipline. Galperin deploys de Certeau's paradigm in his own brilliant study of Austen's readability in something he terms "the historical Austen." I am more concerned with the materials of everyday life, and while I find Thompson's and Galperin's accounts of Austen compelling, I want to introduce another dimension to the economic and material registers of meaning and think about the commerce of writing and the problem of gender and intellectual property in the business of everyday life.[24] Following the alchemical trajectory, female intellectual labor is harnessed in the same way that Indian techne is appropriated and sublimated into a masculinist imperium.[25]

The Labor of Letters

Thus far I have used the examples of plasters and paper to situate the various subject positions for women's labor, intellectual or otherwise. Harriet Smith stands as a fixed embodiment of the material, capable only of literal reading principally because she has no signifying patronym although her spending capacity, anonymously donated, could certainly command a significant dowry. "Portionless" Miss Taylor's interpretative power is largely governed by those she aims to please. Even the limited intellectual freedom she may have had as a governess evaporates when Mr. Knightley pronounces her "very fit for a wife, but not at all for a governess" (38). Her "desperation," as Galperin notes, furnishes her with the agency to secure a husband for herself and to enter the novel in this comfortable position but at a price: the "quasi-servitude" of Hartfield follows her to Randalls where matrimonial and maternal duties subscribe her dependence and she does "as [she] were bid" (38).[26] Emma alone wields both name and portion, and, as if reflecting this privilege, her analytical "skill" enables her to appraise her own condition, dispassionately or not, and at least court the idea of remaining independent of marital ties. Nancy Armstrong, among others, calls into question Emma's vaunted reading skills, noting her many social misreadings, and argues that these shifts from gender to class signify a crucial shift of the "strategic intention of the novel."[27] Although I agree with Armstrong's argument, I also think that in some ways, it doesn't matter if Emma is a skilled interpreter or not because the mere fact that she can imagine a place for herself uncircumscribed by marital law is remarkable. The calculus she uses to fashion such a place is worth noting as well: asked by what means she would occupy herself, Emma replies

> Woman's usual occupations of eye and hand and mind will be open to me then, as they are now; or with no important variation. If I draw less, I shall read more; if I give up music, I shall take to carpet-work. As for the objects of interest, objects for the affections . . . I shall be very well off, with all the children of a sister I love so much, to care about. There will be enough of them, in all probability, to supply every sensation that declining life can need. There will be enough for every hope and every fear; and though my attachment to none can equal that of a parent, it suits my ideas of comfort better than that which is warmer and blinder. My nephews and nieces!—I shall often have a niece with me. (92)

Emma doesn't make any radical deviation from the "usual occupations" available to women, but she leaves men out of the picture. There is no mention of her father or brother-in-law and, more critically, her enjoyment of displaced maternity leaves her quite literally free from the bodily "confinement" of maternity and its consequent blindness and out from under the pesky coverture of a husband's name, enabling her to muse on the pleasures of an abstract future.

Harriet Smith, Mrs. Weston, and Emma Woodhouse embody the opportunities available to their class positions. More complicated is the role of Jane Fairfax, a character circumscribed by epistolarity, who first appears in the novel as a letter. Emma's happy hypothetical reflections on her "ideas of comfort" are quickly grounded when Harriet seizes upon Emma's last exclamation with a literal question: "Do you know Miss Bates's niece? That is, I know you must have seen her a hundred times—but are you acquainted?" (92). Emma's response uncovers the forms of domestic submission she has conveniently removed from her fantasies. She complains how she is "forced to be acquainted" with Jane Fairfax "whenever she comes to Highbury," how her letters are "read forty times over" until "one is sick of the very name" (92). Here, then, the bondage implicit in correspondence, materialized in the paper, appears, *avant les lettres* as it were, to confine Emma in the uncomfortable domestic harness she is made to endure for at least the length of her visit. The second volume opens with Emma calculating whether or not she and Harriet would be "quite safe from any letter from Jane Fairfax," and, at the end of her visit with the Bateses, thinking "that though much had been forced on her against her will, though she had in fact heard the whole substance of Jane Fairfax's letter, she had been able to escape the letter itself" (173).

Galperin identifies *Emma*'s narrative drives in *Emma*:

> Where previously the omniscient style of narration represented the antithesis of epistolarity in matters ranging from control of character to the guidance that this practice exerts of readerly response, Austen appears intent here on exploring and exploiting the affinity of the two narrative modes. This affinity—one that to my knowledge no commentator has yet noticed in *Emma*—resides in the fact that narrative, however plotted *avant la lettre*, may be turned, following the very practice of writing, to the event itself, of which the written (here, as in a letter) is scarcely more than a recapitulation.

The most spectacular feature this affinity yields, argues Galperin, is the suppression of Frank Churchill and Jane Fairfax's courtship narrative. The place of everyday material in relation to the features of courtship might clarify or at least contextualize this Austenian "lapse."[28] The letter that Emma is so eager to evade is first hidden by Miss Bates's homely "huswife" and then by Emma's own forced comments on Jane's handwriting and questions about the Campbells and Dixons until "an ingenious and animating suspicion" nudges her to proceed with an "insidious design of further discovery" (170–171). This design has nothing to do with discovering any hidden truth and everything to do with Emma's risqué fabrication that engages Jane Fairfax's affections to Mr. Dixon's; she has effectively escaped the annoying confines of the "letter itself" and substituted its substance with her own material.

Just as Miss Taylor's courtship with Mr. Churchill happens *avant les lettres* in the novel's prehistory, and just as Harriet's "courtship" with Mr. Elton vanishes in an indignant puff leaving only the maudlin relics of her self-deception, so Jane Fairfax's courtship with Frank Churchill exists in a correspondence marginal to the narrative, only articulated after the fact by his mysterious comings and goings and by her furtive forays to the post office. Elton's charade reveals the twinned allegories of courtship, first, as "the wealth and pomp of kings" and then as "the monarch of the seas," the second historically supplying some part of the first (76). Certainly the courts of the eighteenth century relied on shiploads of Beluga sturgeon to provide their ladies with cosmetic plasters, and when cod replaced sturgeon as counter replaced court, ships were nevertheless deployed. Courts, according to the *Oxford English Dictionary*, are clear spaces "enclosed by walls or surrounded by buildings," yards surrounding castles or, more surprisingly, "formerly also a farm-yard, poultry yard." Such definitions invoke other forms of confinement that include the massive Enclosure Acts passed between 1750 and 1860 that restricted access to common grazing grounds and the resources those grounds provided to villagers. *Emma*'s Highbury enforces a form of domestic enclosure that situates its subjects according to some state of subordination.[29] Locked out of the narrative proper, heir to the court of landed gentry because of her class background and yet faced with an uncertain fiscal future, Jane Fairfax's only discernible means of courtship is through correspondence conducted, presumably, on the cheaper, more widely available paper milled from hemp. Firmly situated in the center of the narrative, Harriet Smith's essay into the world of gentrified courtship relegates her to material marginalia. Like the devolution of court plasters, her only real chance at courtship is with the yeomanry of Highbury, and, appropriately enough, Robert Martin takes her up once again and situates her in her proper court: the farm-yard.[30]

Courtship obviously derives from courtly behavior, but courtship also means an office in the court, a position as a courtier. It is this office that complicates Jane's role in the novel. Midway through the novel, the intellectually valuable but materially bankrupt Jane Fairfax has a short exchange with the socially and culturally vulgar but plainly wealthy Augusta Elton:

[*Jane Fairfax*]: "There are places in town, offices, where inquiry would soon produce something—Offices for the sale—not quite of human flesh—but of human intellect."

[*Mrs. Elton*]: "Oh! my dear, human flesh! You quite shock me; if you mean a fling at the slave trade, I assure you Mr. Suckling was always rather a friend to the abolition."

[*Jane Fairfax*]: "I did not mean, I was not thinking of the slave-trade . . . governess-trade, I assure you, was all that I had in view; widely different certainly as to the guilt of those who carry it on; but as to the greater misery of the victims, I do not know where it lies. But I only mean to say that there are advertising offices, and that by applying to them I should have no doubt of very soon meeting with something that would do." (325)

What I find odd about this conversation is the way in which both characters read the body.[31] Jane Fairfax's initial remark draws attention to the materialization and commodification of an abstract concept: the "human intellect." Mrs. Elton only hears Fairfax's analogy to "human flesh" as literal and immediately conjures up visions of an active and ongoing slave trade although, by the publication of this novel in 1816, this economic practice, in Britain, had been officially abolished for nine years (although the abolition of slavery didn't happen until 1833).[32] Jane Fairfax's response, intended as a clarification of her former comment, mystifies the moment even more. While clearly discounting Mrs. Elton's literal misreading, she nevertheless makes a rhetorical move to endorse the analogy of human traffic: not the "slave-trade" per se but an analogical "governess-trade" that is embedded in an equally murky history of misery and guilt. In this moment of rendering material an abstract concept, "human intellect," material, Jane draws attention to a problem in the marketplace of intellectual labor: that the exploitation of bodies invariably engages the exploitation of minds. In a reversal of alchemy, the notion of paid labor, embedded in Jane's reference to the offices advertising posts for governesses and surely not an entirely foreign concept to the newly propertied, is more foreign to Mrs. Elton than the idea of slave labor which, because of her associations with Bristol and the hints at her fortune's origins in chattel slavery, is probably not that unusual. Mrs. Elton, however, insists on the allusion to slavery even while confronting Jane's looming paid servitude. This conversational moment reveals both a fetishistic disavowal of paid labor while simultaneously uncovering a clear recognition that intellectual labor is worth something: so many rooms, so much "mixing" with family, so much personal comfort per accomplishment. Jane's ruminations are uttered during a conversation in which Mrs. Elton is determined to secure for Jane an appropriate office: a governess of the "*first* situation" (324). Unknown to her is Jane's knowledge of another office available to her—Frank Churchill's courtship. Jane's replies to

Augusta Elton are edged with an irony no reader could understand until the end: when Jane assures her that "a gentleman's family is all that I should condition for," we can read this as a hopeful conclusion to her secret courtship only after we have finished the novel (325).

Paid labor has provided the groundwork for their exchange. Jane Fairfax's surreptitious visits to the post office become the agenda for social dialogue at a dinner party Emma and her father have for the Eltons and Jane Fairfax. Having devoted himself to the "business of being agreeable," John Knightley expresses concern over Jane's morning walk to the post office in the rain. Jane counters this with her desire for exercise and the anticipatory pleasures of receiving letters the post office promises. Knightley teases her over her youthful naïveté, suggesting, that of the two kinds of letters discussed, business and friendship, the latter are worse because " 'Business, you know, may bring money, but friendship hardly ever does' " (317). Jane responds quite passionately:

> Ah! you are not serious now, I know Mr. John Knightley too well—I am very sure he understands the value of friendship as well as any body. I can easily believe that letters are very little to you, much less than to me, but it is not your being ten years older than myself which makes the difference, it is not age, but situation. You have every body dearest to you always at hand, I, probably, never shall again; and therefore till I have outlived all my affections, a post-office, I think, must always have power to draw me out, in worse weather than to-day.(317).

Jane's response to John Knightley's careless remarks is somewhat mystifying, given the putative lightness of their banter. Knightley clarifies his meaning, and suggests that he "meant to imply the change of situation which time usually brings" and adds "ten years hence [Jane] may have as any concentrated objects as I have" (317). Knightley's intent, the narrator assures us, is "kindly . . . and very far from giving offence. A pleasant 'thank you' seemed meant to laugh it off, but a blush, a quivering lip, a tear in the eye, shewed that it was felt beyond a laugh" (317–318).

Having tried somewhat unsuccessfully to ward off the attention to her health and re-establish the necessity for her daily excursion to the post-office, this revelatory moment where her body betrays its feeling, situates Jane in between offices. As paid companion, she risks losing "every body dearest" to her, something that becoming a wife would avert. But as a secret fiancée, the terms of her engagement stretch out to an unforeseeable future, and such an office that should warrant being "in the midst of every dearest connection" remains in the dim margins of the narrative: we as readers have no sure way of understanding her dilemma.

Ever mindful of her new position as a proper member of Highbury's polite society, Augusta Elton translates Jane's walk in the rain into terms of her own understanding. "The man who fetches our letters every day (one of our men, I forget his name) shall inquire for your's too" she announces, adding, "it is a kindness to employ our men" and, over Jane's protests, that "the thing is determined" (319). Her insistent reference to paid labor threatens to sweep Jane into the same anonymous category which, she later does, apparently being incapable of imagining any other future for the "portionless" Jane. Like Harriet Smith, Augusta Elton is governed by the literal and despite her arch allusions to her "lord and master" or "*cara sposa*," she is bound to his "concurrence" (319). Unlike either of these characters and even unlike Emma, Jane appreciates the diverse nuances of female labor, and her ties to the post office are similarly overdetermined. Jane momentarily deflects the scrutiny of her physical health and body by focusing attention on the institution, but in so doing, she aligns the abstraction of epistolary discourse with the physical presence of the body. John Knightley ups the ante (and in more ways than one, because he is also well aware of the material capacities Jane Fairfax offers in terms of marriage dowries) when Jane speculates on the ways in which the postal system works so seamlessly, especially given the "variety of hands, and of bad hands too, that are to be deciphered" (320). John Knightley replies: "The clerks . . . . are paid for it. That is the key to a great deal of capacity. The public pays and must be served well" (320).

Like another invocation of *verba volant scripta manent*, the words that fly back and forth in this dialogue address the material relation between abstract discourse and its residues. The physical letter—a concrete testament to abstract words—represents, for Jane Fairfax, a material relation to absent loved ones; for John Knightley, surrounded by those bodies, the physical letter brings the abstract promise of more property. The exchange of letters is sustained by the paid labor of postal clerks who read, decipher, and deliver these material remainders of value. This long discussion of letter writing leads Jane Fairfax to muse on other forms of the consumption of labor. The writer's "hand," for example, functioning as a metonymy of manual and intellectual labor, also engages Austen's writerly hand that seems to muse on how epistolary discourse, the reigning paradigm for the eighteenth-century novel, was only now recognized as a form of labor. Writing on paper embedded with the material remainders of bondage but understanding the ways in which these residues connect, Jane Fairfax remains bound to the post office where her surreptitious romance—and her ticket out of the miserable life of the paid companion—is hidden. Writing on similar paper, Jane Austen disseminates the shades of English domesticity that, consciously or not, is informed by the subjects of its far-flung empire, while remaining putatively bound to the common room, shielding her labor from prying eyes under the cover of a sheet of blotting paper.

The dialogue between Mrs. Elton and Jane Fairfax represents an alchemist moment in late eighteenth-century and early nineteenth-century British literature. It suggests that the movement from the material to the sublime is equally effective when the process is reversed: rendering the abstract into the material. But what's at stake here is more than an economy of materiality. Jane Fairfax draws attention to the notion of commerce and consumption embedded in the slave trade and governess trade alike. And Mrs. Elton concurs; she replies to Jane's demur with a catalogue of the material conditions Jane can expect from her command of certain forms of knowledge:

> I know you, I know you; you would take up with any thing; but I shall be a little more nice, and I am sure the good Campbells will be quite on my side; with your superior talents, you have a right to move in the first circle. Your musical knowledge alone would entitle you to name your own terms, have as many rooms as you like, and mix in the family as much as you chose;—that is—I do not know—if you knew the harp, you might do all that, I am very sure; but you sing as well as you play;—yes, I really believe you might, even without the harp, stipulate for what you chose;—and you must and shall be delightfully, honourably and comfortably settled before the Campbells or I have any rest. (325–326)

Mrs. Elton's computation is a parody of her command of social value, as any eighteenth-century reader familiar with this world would have recognized. But her concluding remark chillingly echoes Emma Woodhouse's assessment of Harriet Smith's value: the "things" that Mr. Elton's association would give Harriet—"consideration, independence, a proper home" and an "intimacy" with Emma—"things" which, of course, are not things at all. Jane Fairfax's commentary is less easily understood. For her, the relationship between intellect and material value isn't necessarily the burlesque of Mrs. Elton's making as her reply—"You may well class the delight, the honour, and the comfort of such a situation together . . . they are pretty sure to be equal"—makes plain (326). The stunning reality that *Emma* reveals, however, is that abstract concepts like consideration, independence, honor, and delight, all of which define domestic propriety, are most readily available to women as material objects. Here then is a model of value James Thompson identifies:

> In terms of our continued comparison between political economy and the novel, political economy comes to be about money and the novel about character, about subjectivity, a realm and topic far removed from value, for characters/individuals/subjects are not supposed to be quantified, calculated, related to one another comparatively against a single (golden) standard. Monetary expressions of value become inappropriate in the novel, especially in the domestic courtship novel, even though this form concentrates on the evaluation or judgment of individual subjects—what makes one individual more desirable than any other.[33]

Thompson's argument historicizes this model, and suggests that the doctrine of separate spheres is one that has been naturalized and folded into the continued story of the modern centered subject, the individual. William Galperin, James Thompson, Nancy Armstrong, and a host of others make powerful arguments about the history of Austen's novels. But the commodification of Enlightenment values of abstract individualism—something that had already taken place with the advent of the Atlantic slave trade—also defines the place for women. Epistemological value may be read quite differently if we attend to the material histories of paradigms of knowledge in the long and wide eighteenth century. The paper, ropes, and sails that were used to transport goods, bodies, letters, and ideas connect the furthest outposts of British imperial presence to the sitting rooms of rural England. For Austen, this connection also represented a form of oppression; her use of words, like ropes, that connect and transfix the semantic to the material reflects an early modern fascination with the materiality of signification.

Fairfax's casual reference to the commodification of intellectual labor seems to draw on a model of transformation defined by the same language of alchemy that inspired Helenus Scott, Robert Barker, J.Z. Holwell, and even Elihu Yale: the sublimation of base substance into a transparent "truth." The exchange of letters between East India Company factors in India and the Court of Directors and the Royal Society in London, along with the acquisition of commodities, shaped Enlightenment epistemology. What implications did this form of commerce have later in the century? The commerce of an Indian techne may, in fact, have laid the groundwork for the conversation Jane Fairfax and Mrs. Elton have in a nineteenth-century British novel of manners. Lieutenant Ironside's proposal to cultivate the use of the sun plant as an analogous substitute for European hemp presaged the dearth of this plant in the early nineteenth century. The meticulous description of paper making in India he offered to the Royal Society for comment implies that this method might be useful to British manufacturers. If letters have, as Galperin suggests, a "nearly allegorical importance" in this novel, then perhaps considering the materials of correspondence—the ways in which abstract thinking is made material—should be part of the equation. The labor entrenched in correspondence from rope to paper, from hand to pen, from letter to recipient offer another reading of this allegory, one that makes the post office a proper "object" for Jane Fairfax's perambulation, is the same labor that informs Jane Austen's writing in which her hand, tutored to a brilliant finish, surreptitiously produces sheet after sheet of a close-written manuscript.

It is no accident that the preponderance of "hands" presides over the conversation preceding Mrs. Elton's vulgar remarks. John Kightley notices "that the same sort of hand-writing often prevails in a family; and where the same master teaches, it is natural enough. But for that reason, I should imagine the likeness must be chiefly confined to the females, for boys have very little teaching after an early age, and scramble into any hand they can get" (320–321).

This assertion occurs after Jane Fairfax has remarked on the problem of "bad hands" and it seems to be a chivalric testament to female teaching, but, of course, this gallantry is quickly undone when Mr. Knightley compares Frank Churchill's hand to a "woman's writing" declaring that it "lacks strength" (321). The weakness implicit in a woman's writing is taught and learned by the same system that places them under patronymic guardianship. The liberty of assuming "any hand" is reserved for men of the world who can safely dismiss the labor of others in favor of their own business, as John Knightley has made abundantly clear. The clarity and beauty of a woman's hand—Emma, Isabella, Mrs. Weston, Harriet, and Jane all write beautifully—may subscribe to something more insidious than the worldly concerns of lords and masters. Their hands lessen the labor of postal workers, and ease their delivery so that Jane can exclaim: "So seldom that any negligence or blunder appears! So seldom that a letter, among the thousands that are passing about the kingdom, is ever carried wrong—not one in a million, I suppose, actually lost!" (320). This hand, however, is taught by the "governess-trade" that Jane likens to slavery whose "greater misery of the victims" she does not "know where it lies" (325). It is an office that unlike John Knightley's, whose business enables him "to have every body dearest to you always at hand," compels its unfortunate subjects to abandon all "affections" as they move from situation to situation. It is an office that forces Miss Taylor into sixteen years of ayah-like servitude; it is an office that places Jane Fairfax in a strange limbo of marginal existence. And surely it is an office that, in part, kept Jane Austen chained to the common room, which clearly is the fantasy her favorite brother wanted to disseminate:

> She became an authoress entirely from taste and inclination. Neither the hope of fame nor profit mixed with her early motives. Most of her works, as before observed, were composed many years previous to their publication. It was with extreme difficulty that her friends, whose partiality she suspected whilst she honoured their judgment, could prevail on her to publish her first work. So much did she shrink from notoriety, that no accumulation of fame would have induced her, had she lived, to affix her name to any productions of her pen. In the bosom of her own family she talked of them freely, thankful for praise,

open to remark, and submissive to criticism. But in public she turned away from any allusion to the character of an authoress. (324–325)[34]

Commonly relegated to the province of female accomplishments, legible hands were associated with the most banal forms of writing—notes, letters, and unfinished manuscripts—for significant writing was often published writing. Henry Thomas Austen wants to emphasize Austen's reluctance to be published, even if he "outed" Austen in reality, claiming her authorship as a product "entirely" of a female education in the sitting room, somehow exonerating her of any charge of moving into circles of literary "notoriety." The hope of attaining a "profit" from her literary labor, blissfully claims Henry Austen, is "unmixed" with her talent: her hand is safely confined to the domestic "bosom."

The geography of the written word, as Miles Ogborn identifies, "offers ways of understanding the relationship between space, knowledge, and power in the practices of European trade and empire that show these practices in the process of their construction and operation."[35] Even if Austen's authorial hand remains confined to her family, *Emma* insists on representing the labor of letters as one that navigates the outposts of empire. C.A. Bayly has argued that Mughal indifference toward the printing press the Portuguese were pressing upon them attested to the power of the scribal tradition, and legible hands were vaunted, valorized, and validated as part of the ideals of kingship.[36] Obviously, print culture, flourishing in early modern England, made its presence known in India by the nineteenth century, yet scribal work remained the most efficient means of communication; in the Company's ledgers, accounts, and correspondence, a legible hand was imperative.[37] Although *Emma*'s dramatic action never leaves Highbury, the sheer amount of work accomplished by letters, written on palpable metaphors of imperial toil, link the sheltered (feminine) life of rural England with the harsh (masculine) experiences of the East India Company's "servants," vulnerable strangers in a strange land.

The British harnessed Indian material and technological substances and methodologies to supply their own scientific wants. Such accruement suggests that somehow these forms of knowledge were extant without human agency. The entrenchment of colonialism as a political force to keep open profitable mercantile exchange privileges the idea that Indians were present, in this process of abjection, only to supply brute labor, to become as inanimate as the materials their culture "offered" their conquerors and as unconnected to the epistemology they contributed to the interests of Western science as farm animals are to agricultural production.

Not one of the correspondents from the East India Company as much as named a single "Hindostan" paper maker, Brahmin inoculator, Madrasi mortar manufacturer, Allahabadan ice maker, or Bengali cloth dyer. To the British, these forms of manufacture and practices of science were as abstract and unreadable as the substances themselves. Nevertheless, this epistemology found itself lodged in the archives of the earliest compilation (in England) of Enlightenment science, the *Philosophical Transactions of the Royal Society*, and not simply in its early years, when Britain's purchase in commercial trade with India was tenuous at best, but throughout the eighteenth and into the nineteenth century after Britain had secured a colonial presence with the defeat of Tipu of Seringapatam, the Sultan of Mysore in the 1790s. These letters and reports almost undoubtedly informed the shape and trajectory of Western Enlightenment science as the papers on smallpox inoculation demonstrate.

But what the British also sublimated was the Indian physical and intellectual labor that extracts, refines, and disseminates the properties of the substances. The technique and techne that constituted Indian intellectual property were engaged in a continual traffic first by the British stationed in India, and later by scientific societies responsible for the dissemination of those ideas. They resurfaced in the languages of British taxonomy and British epistemology, demonstrating how profound the commerce in this kind of property was. The history of British imperial oppression of Indian natives, therefore, may not have simply been organized around the exploitation of bodies, but also of minds, that, in turn, had to be rendered as bodies in order to make their subjugation absolute.

I ended my discussion of the traffic in *Emma* with a reference to the notion of labor in epistolary discourse, whether it is embodied in the paid clerks delivering letters or in the production of the novel. But this discursive form of work didn't stop there. The *Letters Received by the East India Company from Its Servants in the East* make it abundantly clear that letter writing is a form of labor, if only in the abstracted pursuit of science. It seems as if the conversation in the novel is possible only if we take its materiality into consideration. The exchange between Jane Fairfax and Mrs. Elton is comfortably regurgitated in the drawing room after a dinner prepared by other paid bodies. Austen's text, presumably created sometime in the early teens of the nineteenth century, depicts a solidified culture of manners in England, informed by England's growing status as an imperial power. That culture is on the brink of understanding its role as an imperial power defined by an abstract humanism. No longer gathering of commercial and epistemological material, Britons now imagined themselves endowing their imperial world with "civilization." But it would not have been possible were it not for an earlier understanding on the part of the British that the other represented not simply a body to conquer but a body of discourse that mattered, and therefore had to be mastered.

In 1772, Sir Robert Barker made a journey to Benares to confirm the stories he had heard that ancient Brahmins had "a knowledge of astronomy." In a paper that he subsequently sent to the Royal Society—and that was published later in 1777—Barker describes an ancient observatory of considerable dimension:

> a large terrace, where, to my surprize and satisfaction, I saw a number of instruments yet remaining, in the greatest preservation, stupendously large, immoveable from the spot, and built of stone, some of them upwards of twenty feet in height; and, although they are said to be erected two hundred years ago, the graduations and divisions on the several arcs appeared as well cut, and as accurately divided, as they had been the performance of a modern artist.[1]

Eager to learn of a Brahmin method of calculating an approaching eclipse, Barker demonstrates the same kind of wonder in observing the techne of

Brahmin astronomy as he does of the methods of his ice-maker in Allahabad, even putting to rest "some doubts . . . with regard to the certainty of the ancient Brahmins having a knowledge of astronomy" by definitively claiming that "we know that the present Bramins pronounce, from the records and tables which have been handed down to them by their forefathers, the approach of the eclipse of the Sun and the Moon, and regularly as they advance give timely information to the emperor and the princes in whose dominion they reside."[2] The difference in this letter from the ones I've discussed is in the immobility of this form of techne. That is, mortar, ice, smallpox inoculation, paper, and textiles all share a property of mobility. Isaac Pyke's recipe includes English substitutions, Barker's excitement at the methods of ice making is partially fueled by his ideas of exploiting this techne at places of British residence, Coult's and Holwell's commitment to the method of inoculation is that it could be usefully adopted in Britain where the disease raged. Ironside offers a solution to cotton prices and the shortage of hemp again by suggesting the cultivation of an indigenous Indian plant in Europe. But Barker's description of Brahmin astronomy remains oddly fixed within the confines of the observatory at Benares.

Thinking through the geography of imperial technology, Simon Werrett suggests that the emphasis historians place on local knowledge is accomplishing precisely what it critiques about universal macro-narratives: replacing one account with another only manages to fetishize another standpoint.[3] Instead, he argues that

> understanding the geography of imperial technology . . . demands a sensitivity to the way that both universalist and localist discourses served as actors' categories. By uncovering the work that historical actors did to sustain these positions, we can more clearly understand how these actors attempted to assert their own relevance to technology transfer.[4]

I have been championing the value of local techne and have uncovered the ways in which strategic amnesia allowed for the emergence of a triumphalist European account of their exceptionalism in the history of technologies of colonialism. It is clear, however, that a whole host of actors, to borrow Latour's phrase, are necessary to make narratives of whatever size. Emerging from the myriad methodologies for mapping spaces of alterity, encounter, and conquest is the necessity for having numerous discursive patterns to relay these transfers of knowledge and techne, and I have argued that the deployment of analogy and of alchemy as a paradigm of transfer was critical to East India Company servants trying to make sense of a world they had never encountered. Analogy offers a means by which one could interrogate the relationship between material objects and their social and semiotic power. Analogy also reveals the complexity and richness of the Anglo-Indian encounter, one that navigated unknown chasms through an older form of reason—alchemy—that allowed room for wonder. Interestingly, Barker's visit to Benares and to the monumental observatory, replete with its massive instruments, while eliciting a respect for Brahmin astronomical knowledge, did not provoke the same excitement that his observations of ice making did. That is, Barker did not see the possibilities for exploiting Brahmin methods of calculating eclipses for British use. Even though a mere two years separated Barker's papers on ice making and astronomy, it seems as if a discursive shift was taking place, one that was influenced by the appropriation or transfer of Indian techne into the archives of the Royal Society.

Werrett argues that taking both universalist and localist approaches into account opens up the possibility of a third position, "which emphasizes the important role of *immobility* in the geography of imperial technology."[5] If we look at the difference in the approaches Barker makes toward ice making in Allahabad and the Brahmin's observatory in Benares, we can trace the role of immobility in, what Werrett terms, the geography of imperial technology. Werrett investigates the provenance of rockets that William Congreve claimed to have invented but who also knew about rockets used against the British by Tipu Sultan and Haidar Ali from 1767 to 1799.

> In order to assert maximum control over the design and manufacture of rockets, both Congreve and the East India Company insisted that they should be managed and improved locally before being dispatched for use at a distance. As one company memo urged: "The Rocket should be made on the spot." The question was, which spot?[6]

This is a question I'm rearticulating in order to account for the difference in Barker's responses to Indian techne but also, more broadly, to think about immobile geographies and the production of technologies of colonialism. One of the effects of immobile geographies is that they stabilize forms of knowledge by making their origins clearly rooted in the soil of that particular territory: the "spot" that the East India Company invokes. Of course, these ideas and epistemologies move, as Werrett makes abundantly clear and as I have argued throughout this study, but the ideology of dominance, as in the case of colonialism, makes it imperative that there is a hierarchy of the "spot."

England's mercantile power in the global world was precarious, as the letters from the East India Company from the early seventeenth century demonstrate. Nicholas Downton and Thomas Kerridge both sailed into unchartered waters when it came to negotiating audiences with Mughal emperors. Two events occurred in the first decade of the eighteenth century that changed the shape of England's national identity and paved the way for more accurate trade routes that, ironically, led to the ideologies of immobile geographies. The first was the 1707 Act of Union that annexed Scotland to England and Wales to form a single united kingdom of Great Britain and, in the process, created the first Parliament of Great Britain. The second was the naval disaster that occurred six months later when Clowdisley Shovell miscalculated his location and scuttled his ships on the treacherous rocks of the Isles of Scilly, an incident that helped launch the 1714 Longitude Act.[7]

But Shovell's naval disaster and the formation of the United Kingdom have more in common other than occurring in the same year. These events are a crucial intersection between material geography and the psychosocial formation of national identity that link trauma, trade, and the "global" position Britain navigated in the long eighteenth century. Considering the first two terms, it might be useful to recall Cathy Caruth's contention concerning trauma and the possibility of history, where she argues that in the "bewildering encounter with trauma . . . we can begin to recognize the possibility of a history which is no longer straightforwardly referential."[8] We also have to remember the state of Britain's global navigations; Shovell's 1707 disaster may well have portended the demise—or at least delay—of a share of the lucrative West Indies trade, especially during a time when England's purchase in India was anything but stable or dependable.

While this may seem obvious, there is a significant difference between latitude and longitude; the zero-degree parallel of latitude is fixed by the equator, while the zero-degree meridian of longitude can shift according to the political weather.[9] Gauging one's latitude on the open seas is not a difficult matter because of the stability of length of day, height of sun, or guide stars above the horizon. But determining one's longitude on a plane with few visibly reliable markers is a different matter, and charting one's exact position on that plane, involves the need for a stable third term—in this case, the precise knowledge of time aboard ship and time at the home port (or other place of known longitude). Even a few seconds can make a drastic difference (as Shovell found out), and the rolling of the ship or shifts in barometric pressure can all contribute to changing a clock's accuracy. Early modern European transoceanic travel was, therefore, confined to a few narrow shipping lanes shared by ships of all seafaring nations, including pirates. Such were the terrors of uncharted waters that ships radically hostile to one another would risk capture even while they looked for secret routes to avoid costly confrontations.

One navigational issue is the difference between latitude and longitude; the former requires a binary relation between a ship and a fixed point (sun, stars, horizon, length of day, etc.) while the other is complicated by the need for a stable third term in order to yield a reliable account of where one was on *mare incognita*. John Harrison is largely credited with solving the problem of charting longitude with his invention of the marine chronometer that kept remarkably accurate time. But lunar distance calculus was used with greater frequency until it was replaced by wireless telegraph signals in the twentieth century.[10] These signals were, in turn, replaced by space-based satellites that constituted the Global Positioning System of Roger Easton's design, a navigational principle that became fully operational in 1994 and has subsequently been integrated into the common parlance of a new millennium.

The Global Positioning System is hardly new. It addresses a problem that any early modern sailor would be familiar with: how do we know where we are? Although maps were quite reliable in identifying where one had been and where one was going, even Mercator projections couldn't really tell a ship on the high seas where they were. Maps are binary systems; they allow one to calculate from one fixed point to another in the same ways that latitudinal calculations can guide ships from one point to another as Columbus did on his 1492 voyage across the Atlantic. But English ships trying to avoid the Dutch or the Portuguese on their quests for nutmegs had a more difficult time. Even if they didn't encounter pirates, miscalculations of their course could be disastrous, and sources of fresh water and food could elude them for months and plague a crew with the horrors of starvation, scurvy, bloody flux, and the like until there were hardly any people left to sail the ship. One needs a form of trilateration in order to determine a reliable sense of place, a stable sense of identity, and secure cargoes.[11]

Earlier I suggested that connections between trauma, trade, and global position may be read through an intersection of material geography and the psychosocial formation of nation. I want to return to that with the idea of trilateration in mind. Just as the third term/position is imperative to establishing one's exact situation—*where* one is—so is the third term vital to determining precisely *who* one is. The 1707 Act of Union brought together two powerful Parliaments, having always been united, it is true, under one monarch, but uneasily so as neither side quite trusted the other not to turn Catholic (for example) or engage in any number of rebellions. Their consolidation, however, was contingent on a third term—Ireland—who had no voice in the Union even after pleading for representation. Thus this "other"—an "absolute" as Gayatri Spivak once formulated—guaranteed the solid footing for Great Britain to sally forth, as it did in 1713, one of the victors in the War of Spanish Succession. In other words, in order for England to have a global position, it was important to stabilize its uneasy relation with Scotland, and the 1707 Act of Union performed a moment of self-consolidation by abjecting Ireland. The psychosocial formation of nation preceded the material evidence of imperial power, but it was necessary to have this paradigm in place in order to position oneself as part of a global economy. England's relation vis-à-vis Turkey, for example, was one that was mediated through France as writers like Daniel Defoe exploited in one of his earliest publications, *An Account of Monsieur De Quesne's Late Expedition at Chio Together with the Negotiation of Monsieur Guilleragues, the French Ambassador at the Port/ in a Letter Written by an Officer of the Grand Vizir's to a Pacha; translated into English* (1683). This fictional account is organized around a trilateral dynamic. The author dedicates his account to the Right Honourable George Lord Marquess of Hallifax Lord Privy-Seal, &tc. (whose complicated relationship to James II probably endeared him to Defoe) because his "Services to the Crown, and Merits from the Nation are so great, that Time and the Memory of them must be of equal Durance." To the reader, Defoe declares that "meeting with the letter in *French*, have made the Sense of it *English* . . . I publish it, as well to shew the Pride and Insolence of Humane Nature, when Ignorance is possest of Absolute Power, as the Dissimulation, Fraud, and Corruption of any Sect, who pretend to be God's Elect, or only chosen People, which the *Mumelans* do."[12]

Perhaps most telling is Alexander Dalrymple's account of the existence of *Terra Australis Incognita*. Geographer, hydrographer, and writer for the British East India Company, Dalrymple's translations of Spanish documents captured in the Philippines in 1752 lead him to publish *An Account of the Discoveries Made in the South Pacific Ocean Previous to 1764*. In this account, he describes himself as having gone to the "East-Indies in the service of the Company, at an early age," and being "inflamed with the ambition to do *something* to promote the general benefit of mankind, at the same time that it should add to the glory and interest of [my] country." His proposal is careful to note that because no "observations of longitude having hitherto been made, to determine the situation of any of these places, or even to regulate the western limit, it is far from being imagined that their situations attain a minute precision."[13] Writing from Madras, consolidating his translation of Luis Váez de Torres's record of sailing the strait separating Australia from New Guinea, Dalrymple deploys a trilaterated national identity in order to give a more minute precision to the global position of this unknown territory, claiming its knowledge for the imperial glory of Great Britain. Shortly after Dalrymple published his account, states Dava Sobel, "captains of the East India Company and the Royal Navy flocked to the chronometer factories," adding that in 1791 the East India Ccompany "issues new logbooks to the captains of its commercial vessels with preprinted pages that contained a special column for 'longitude by Chronometer.'"[14] Britain had a made a name and a place for itself by charting the seas.

British national identity was fashioned through trilateration that involved not only other European nations vis-à-vis the radical alterity of Asian and

African ones, but also itself. The psychosocial structure of Great Britain provided a paradigm against which one could come to terms with difference, but it also accommodated the possibility of exchange. That is, not simply as a binary structure—like "the West and the rest"—trilateration allowed for the entry of different epistemological genealogies into British histories. To think about a system of global positioning is to rethink a straightforwardly referential history, as Caruth claims trauma forces us to do; to permit history to arise when immediate understanding fails. The Enlightenment was not shaped primarily by ideas circulated among Europeans and Britons nor was it shaped primarily by nations outside this circle but by a dynamic whose stability depended on three points of contact. To return to the notion of immobile geographies, while the prime meridian is of vital importance because it allowed Britons to imagine a chart of the oceans as national territory, it had to be imagined as part of an immobile geography rather than the referential phenomenon that it is in reality. Nevile Maskelyne published 49 issues of the *Nautical Almanac* based on the meridian at the Royal Observatory in Greenwich. His tables "not only made the lunar method practicable, they also made the Greenwich meridian the universal reference point. Even the French translations of the *Nautical Almanac* retained Maskelyne's calculations from Greenwich—in spite of the fact that every other table in the *Connaisance des Temps* considered the Paris meridian as the prime."[15] Thus the center for longitude was fixed in the center of imperial Britain, with a whole history of meridians imagined by Eratosthones and Ptolemy, Mercator and the Cardinal Richelieu rendered meaningless, much less other influential Chinese, Indian, Persian, or other geographers that have been abjected from this history. Nations call on similar ideologies to render an imagined community material.[16] Having secured a stable method of charting oceans toward the latter half of the eighteenth century, Britons could now start projecting Enlightenment notions of inert matter onto colonial topographies and render those geographies and their techne immobile, subject to an empirical analysis that eventually resulted in nineteenth-century narratives of the technological paucity in India. But, as I argued at the conclusion of chapter 5, the ideology of immobility was necessary because of the extraordinary motility of putatively abject materials, whose capacity for spontaneous and active movement consumed the energies of early modern servants of the East India Company. Writers like Volney did lament the "fall" of past empires; yet they had decided that despite its antediluvian history of civilization, the East was intellectually and morally bankrupt.

> Here . . . once flourished an opulent city; here was the seat of a powerful empire. Yes! these places now so wild and desolate, were once animated by a living multitude; a busy crowd thronged in these streets, now so solitary. Within these walls, where now reigns the silence of death, the noise of the arts, and the shouts of joy and festivity incessantly resounded; these piles of marble were regular palaces; these fallen columns adorned the majesty of temples; these ruined galleries surrounded public places. Here assembled a numerous people for the sacred duties of their religion, and the anxious cares of their subsistence; here industry, parent of enjoyments, collected the riches of all climes, and the purple of Tyre was exchanged for the precious thread of Serica, the soft tissues of Cassimere for the sumptuous tapestry of Lydia; the amber of the Baltic for the pearls and perfumes of Arabia; the gold of Ophir for the tin of Thule.[17]

This melancholy reverie is followed by an explication of the "law of nature" that, for Volney, favored the inauguration of an Enlightenment republicanism and therefore situated Europe as the underwriter of a prosperity in perpetuity. Missing from Volney's romanticized account is John Locke's "rule" of analogy as the critical method by which one can link old and new epistemologies, and observed and unobserved phenomena.[18] What emerges instead is a dominant Whiggish history that valued empirical analysis and, as Mary Poovey has argued, the triumph of the modern fact that permanently tied "certain critical fictions" of money and price to systematic knowledge.[19] These powerful epistemological networks were crucial to the production of a narrative of colonial dependence, even while Britain's imperial geography was utterly contingent on the disavowal of alien techne.

ACKNOWLEDGMENTS

This book has taken me a lifetime to write. Starting with my parents' immigration to the United States about a decade after Indian independence, my peregrinations between the United States and India and their widely divergent cultural standpoints over the past fifty years have altered my understanding of what is "strange" and "exotic." I have had a great deal of help from my family in producing this book, and because it has taken me so long to write, I have lost beloved members whose sustenance and love means more to me than I could ever describe: my father, Ravindra Nath Sudan; my uncle Sanjoy Lahiri; and my "Mashi-Pishi," Karuna Mahtab. To them I owe a debt of utter gratitude, as well as to my mother, Dipali Sudan, and my uncle and aunt Danny and Nandini Mahtab, who have made it possible for me to do work in the Indian archives. Heartfelt gratitude to my brother, Ranjeet, with whom I traveled to Benares to cast my father's ashes into the Ganges; together with my sister-in-law, Melissa, and my remarkable nephew and niece, Anil and Anjali. I am fortunate to have this wonderful family.

I have benefited from grants and fellowships from Southern Methodist University and the Folger Shakespeare Library to conduct research at the Folger, the British Library, the Asiatic Society and the Victoria Memorial Society in Calcutta, the Rajah Murthia Library in Madras, and the National Archives of India in New Delhi. I very much appreciate the support that these institutions have given me. I presented some of the chapters at invited lectures and colloquia at the University of Rochester, Rice University, the University of Pittsburgh, Stanford University, North Carolina State University, Texas Tech University, the University of Texas at Arlington, the Asiatic Society and the Victoria Memorial Society in Calcutta, and the University of Kent in the UK.

I want to thank members of my department for their insights and the humor with which they often entertained my project: Rick Bozorth, Tim Cassedy, Darryl Dickson-Carr, Dennis Foster, Bruce Levy, Dan Moss, Beth Newman, Tim Rosendale, Jayson Sae-Saue, and Nina Schwartz. To my colleague Lisa Siraganian I owe more than I can say: her support has been my mainstay during the final months of the book's production. I want to thank the members of the Dallas Area Social History group for listening to earlier versions of various chapters and responding as the fine historians they are; I've learned a lot from these exchanges. My long relationships with Stacy Alaimo, Adam Brenner, Tom Pribyl, and, more recently, Jenni Brakey and my amazing coeditor of *Configurations*, Melissa Littlefield, have sustained me through many hours, bright and dark.

Helen Tartar was one of the first editors to express interest in publishing this manuscript whose borders were undefined by any one discipline and whose interdisciplinary moves made many people question what kind of book this would be. I owe her a tremendous debt for the faith she had in my work, and her untimely death has made the world of academic publication a poorer place. Thomas Lay has been an inspiring editor, and I thank him for his patience, insight, and editorial brilliance.

Others in the profession have listened many times to my presentations, and their incisive critique has made this a better book. Srinivas Aravamudan, Mohammed Bamyeh, Jill Casid, Tita Chico, Jay Clayton, Billy Galperin, Humberto Garcia, Randall Halle, Gina Jenkins, Olivera Jokic, Nicole Jordan, Suvir Kaul, Ann Louise Kibbie, Mimi Kim, Jonathan Lamb, Devoney Looser, Ruth Mack, Bridget Orr, Adela Pinch, Laura Rosenthal, Alvin Snider, James Thompson, Robert Travers, Marlene Tromp, Dan Vitkus, Chi-Ming Yang, and the late—and sorely missed—Hans Turley have all helped me think through some of the earlier problems. There are probably others whom I haven't mentioned to whom I offer my thanks and apologies for a failing memory.

Donna Landry and Gerald Maclean continue to be among my most favorite thinkers; their work has encouraged my own through the years. Betty Joseph read an earlier version of this manuscript, and I'm deeply

grateful to her for the enormously helpful session with the Early Modern Reading Group to which she kindly invited me to speak and for helping me see just what I was arguing. I am indebted to Ronald Schleifer for some of the most breathtakingly precise and detailed commentary I have ever received.

To Bob Markley I owe far too much. His work on the intersections between China, Japan, India, and the English imagination was the inspiration for my own thinking; I thank him and Lucinda Cole for their generosity and for our ongoing conversations. Special thanks to Sharon Willis for her brilliant editing of the introduction and for telling me that the book was done, which is harder to know than one would think. She and John Michael have been wonderfully supportive friends in all matters for the past thirty years. If it were not for Pat Gill's discourse on "reigning paradigms," I doubt I would have pursued alchemy to the extent that I did. She and Richard Wheeler continue to be dear friends and my greatest reality checks. Tom DiPiero's critique of reason, rhetoric, and the importance of form, of course, deeply influenced my thinking. I thank him for the years of intellectual exchange and for his friendship.

I thank Farley and Emily Morris for their hilarity and wit and for the many (sometimes heated) conversations and performances we've shared. My greatest debt, however, is to my most challenging critic, Chris Morris, who taught me to think differently and pay attention to archives. He has believed in my project from its inception, and this book is as much his as it is mine. I hope that doesn't come back to haunt him.

INTRODUCTION: MUD, MORTAR, AND EMPIRE

1. I use the term "techne" to refer to the "art, skill, or craft; a technique, principle, or method by which something is achieved or created," as defined by the *Oxford English Dictionary*. However, I am fully aware of the confusion that this term may cause, given its philosophical association with *episteme* (and Aristotle and Heidegger). Still, I insist on using this term to deconstruct its European provenance, and to demonstrate rhetorically the complicated knowledge bases of European techne.

2. The nexus of this crisis is the reconceptualization of value from, as James Thompson argues, "treasure to capital and the consequent refiguration of money from specie to paper." Thompson's argument follows the complicated path to identify the various problems with value and representation that characterized the long eighteenth century. See James Thompson, *Models of Value: Eighteenth-Century Political Economy and the Novel* (Durham and London: Duke University Press, 1996), 2.

3. *Philosophical Transactions* 37: 233.

4. *Philosophical Transactions* 42: 256–257.

5. *Philosophical Transactions* 30: 225. An interesting note on Father Papin's observations of Indian artisans: he focuses on the success with which these artisans elude the efforts of (presumably) European visitors to find fault with Indian textiles, which presumes an act of rivalry or challenge that underwrites the Jesuit mission.

6. Daniel Defoe, *Everybody's Business Is Nobody's Business*, Project Gutenberg, http://www.gutenberg.org/ebooks/2052.

7. Even as late as 1798, Jane Austen makes a reference to "true Indian muslin" in *Northanger Abbey*; Henry Tilney demonstrates his knowledge of textiles and their quality by alluding to the difference a "true" Indian muslin makes. For a fine reading of this moment, see Lauren Miskin's "'True Indian

Muslin' and the Politics of Consumption in Jane Austen's *Northanger Abbey*," *Journal for Early Modern Cultural Studies* 15:2 (Spring 2015): 5–26. For an excellent account of the textile crisis in England, see Chloe Wigston Smith, "'Callico Madams': Servants, Consumption, and the Calico Crisis," *Eighteenth-Century Life* 31:2 (Spring 2007): 29–55.

8. *Philosophical Transactions* 22: 225–226.

9. Ursula Klein and Wolfgang Lefèvre, *Materials in Eighteenth-Century Science: A Historical Ontology* (Cambridge, MA: MIT Press, 2007), 7.

10. Quoted in Christopher Hill, *Century of Revolution, 1603–1714*, Vol. 5 (Edinburgh: T. Nelson, 1961), 92.

11. Klein and Lefèvre, *Materials in Eighteenth-Century Science*, 9.

12. Ibid., 19–20.

13. I realize that there are differences between the reception of observations between France and England, between Defoe and his investment in trade and the order of Jesuits and their pursuit of more intellectual interests, but the point I'm making is that travelers from England and Europe were alike in their admiration of Indian techne.

14. Robert Travers, for example, identifies the ways in which old visions of a powerful and cohesive British nation confronting a weakened and divided India were replaced by a picture of British traders forging strategic alliances with Indian capitalists. Increasingly, the causes of British expansion were sought as much in indigenous processes of change like the "commercialization of power" and the drift of "intermediary groups" toward the East India Company, as in endogenous factors like the growth of British power or ambition. Although Travers suggests that debates within recent historiography concerning issues of Indian agency versus cultural dislocation contest the totalizing capacities of either position, he also argues "both strands together have done much to uncover the complexity of early modern India from the narrowness and distortion of older imperialist accounts." See Robert Travers, *Ideology and Empire in Eighteenth-Century India: The British in Bengal* (New York and Cambridge: Cambridge University Press, 2007), 12.

15. One must pay tribute to Stephen Greenblatt in any discussion of marvels. It is less the marvels themselves that I am interested but, rather, the model of history making he identifies with encounter narratives. Functioning as part of a "culture's representational technology," Greenblatt locates the anecdote between "the undifferentiated succession of local moments and a larger strategy toward which they can only gesture." The East India Company letters are not intended as anecdotes; they are reports that range from early seventeenth-century inventories and ship logs to early nineteenth-century medical reports. They do, however, function similarly in terms of

contributing to the "larger progress or pattern that is the proper subject of a history perennially deferred in the traveler's relation of further anecdotes." See Stephen Greenblatt, *Marvelous Possessions: The Wonder of the New World* (Chicago: University of Chicago Press, 1991), 3.

16. Bruno Latour, *Pandora's Hope: Essays on the Reality of Science Studies* (Cambridge and London: Harvard University Press, 1999), 151.

17. See John Barrell, *The Infection of Thomas De Quincey* (New Haven: Yale University Press, 1991); Nigel Leask, *British Romantic Writers and the East: Anxieties of Empire* Cambridge); Saree Makdisi, *Romantic Imperialism: Universal Empire and the Culture of Modernity* (Cambridge: Cambridge University Press, 1998); and Charles J. Rzepka, *Sacramental Commodities: Gift, Text, and the Sublime in De Quincey* (Amherst: University of Massachusetts Press, 1995).

18. By this I am referring only to the conclusion of this particular nightmare, not the end of the *Confessions*, something that would be much harder to pinpoint given the numerous sequels and appendices De Quincey added to his original memoir.

19. It is interesting that Lytton conflates geographical *topoi* in ways similar to De Quincey.

Lord Lytton to Lady Lytton September 1877, Thomas Macaulay to his sister June 1834, quoted in Kavita Philip, *Civilizing Natures: Race, Resources, and Modernity in Colonial South India* (New Brunswick, NJ: Rutgers University Press, 2004), 29.

20. Philip, *Civilizing Natures*, 30–31. The irony of the Anglicized terrain that Lytton contemplates with such delight, relishing its capacities to conjure England, is that it also reflects the fearful oriental topography that so plagued De Quincey. His *Confessions* famously remark on the Malay wandering about the Lake District, by whose "means [he has been] transported into Asiatic scenes." The cultivation of opium poppies in India by early nineteenth-century compulsory schemes was most prolific in Bihar, a province near the Nepal border and the foothills of the Himalayas. The mountainscape of Ooty would not have looked dissimilar to the Rajgir hills of Bihar, where among the Romantic scene of the Lake District a Malay—and his opium—is foregrounded.

21. Daniel Defoe, *Robinson Crusoe* (New York: Bantam, 1981), 93.

22. Ibid., 108.

23. Virginia Woolf, *The Common Reader* (New York: Harcourt, 1925).

24. I am indebted to Lydia Liu's highly influential reading of Robinson Crusoe's earthenware pot. Liu contends that Defoe's focus on earthenware "at the height of the European craze for true porcelain" is in part due to Defoe's

ambivalence toward the importation of luxury goods. But more crucially, it is Defoe's tale of a "(white) man's *solitary survival in nature*," which reshapes the experiment with earthenware as "symptomatic of . . . the poetics of colonial disavowal." I agree with Liu, but what I also find noteworthy is the mud itself that furnishes the grounds for such a disavowal. See Lydia H. Liu, "Robinson Crusoe's Earthenware Pot," *Critical Inquiry* 25:4 (Summer 1999): 731–733.

25. I am indebted to the wealth of scholarship that attends to Defoe's spatial politics. In particular, critics as widely spaced as Eric Berne and Richard Nash address the issues of environment and sovereignty in ways that complicate traditional receptions of the novel as one of the triumphal articulations of psychological realism. See Eric Berne, "The Psychological Structures of Space with Some Remarks on *Robinson Crusoe*," *Psychological Quarterly* 25 (1956): 549–567; Jonathan Lamb, *Preserving the Self in the South Seas, 1680–1840* (Chicago: University of Chicago Press, 2001); Richard Nash, *Wild Enlightenment: The Borders of Human Identity in the Eighteenth Century* (Charlottesville and London: University of Virginia Press, 2003); and John Richetti, "Robinson Crusoe: The Self as Master," in *Robinson Crusoe: An Authoritative Text, Context, Criticism*, ed. Michael Shinagel (New York: Norton, 1994), 357–373.

26. Defoe, *Robinson Crusoe*, 42. The word "apartment" in this context is illuminating. Originally referring to a portion of a house or building that is allotted for the use of a particular person, in this context, Crusoe's refuge in the "thick bushy tree like a fir" gives agency over to the island itself, and the subsequent "country house" and "seacoast house" he "fancie[s]" himself later possessing can only be possible because of the original portion of island flora allotted to him.

27. Defoe, *Robinson Crusoe*, 138. I realize that the footprint is in sand, not mud, a different medium—although technically belonging to a textural class of soil. Literarily, however, this sand is not the kind of loam that providentially yields the bounty sustaining Crusoe over the years but, rather, offers the irrefutable sign of something other than a providential sovereign. Planted on the border of the island, this footprint disrupts any ground Crusoe has covered, leaving him back to the edges of his island survival (in much the same way that sand itself may be read as a border soil).

28. Defoe, *Robinson Crusoe*, 138–139.

29. By modern forms of imperial dominion, I refer to the complicated and transnational position that wireless discourse occupies. That is, all forms of wireless exchange depend on the ore coltan (columbite-tantalite) to produce tantalum capacitors that are highly resistant to heat and very efficient conduc-

tors. This ore was primarily mined from the mud of the Democratic Republic of Congo, although when word got out about the multifarious levels of exploitation occurring in the mining camps, powerful nations like the United States declared embargoes against the Congo, after securing sources elsewhere, in Brazil. In the case of modern transnational corporate imperialism, mud sublimates into the very sophisticated discourse of wireless technology.

30. Sublimation is also a key term in Freudian psychoanalysis in which potentially harmful drives are channeled into something more socially acceptable, for example, dangerous aggression into athletic competition. In some ways, this form of sublimation also speaks to my argument in that "dangerous" foreign techne is sublimated to the benefit of the British.

31. Alan Bewell, *Romanticism and Colonial Disease* (Cambridge: Cambridge University Press, 1999), 252.

32. Julia Kristeva, *Powers of Horror: An Essay on Abjection*, trans. Leon S. Rodriguez (New York: Columbia University Press, 1982), 8–11.

33. Ibid., 11.

34. I am deeply indebted to Robert Markley whose work has inspired my own study. For an illuminating account of Defoe's novel for changing our ways of seeing and understanding British colonial enterprise, see his *The Far East and the English Imagination: 1600–1730* (Cambridge: Cambridge University Press, 2006).

35. Lady Mary Wortley Montagu to Miss Anne Thistlethwayte, April 1717 in Christopher Pick, ed., *Embassy to Constantinople: The Travels of Lady Mary Wortley Montagu* (New York: New Amsterdam Books, 1988), 125.

36. Lady Mary Wortley Montagu to her sister, Lady Mar March 10, 1718, Pick, 167.

37. Steven Shapin, *A Social History of Truth: Civility and Science in Seventeenth-Century England* (Chicago and London: University of Chicago Press, 1994), xxv.

38. Claudia L. Johnson, *Jane Austen: Women, Politics, and the Novel* (Chicago and London: University of Chicago Press, 1988), 121.

39. Jane Austen, *Emma*, eds. Richard Cronin and Dorothy McMillan (Cambridge: Cambridge University Press, 2005), 325.

40. Maaja A. Stewart, *Domestic Realities and Imperial Fictions: Jane Austen's Novels in Eighteenth-Century Contexts* (Athens and London: University of Georgia Press, 1993), 137.

41. British Library, Oriental and India Office Collection, *Letters Received by the East India Company from Its Servants in the East*, Vol. I: *1602–1613* (London: Sampson, Low, Marston & Company, 1896), Document 110: 277–278. See also Travers, *Ideology and Empire in Eighteenth-Century India*, 22.

42. Ronald Schleifer, *Analogical Thinking: Post-Enlightenment Understanding in Language, Collaboration, and Interpretation* (Ann Arbor: University of Michigan Press, 2000): 4.

43. Elizabeth Ermarth, *Realism and Consensus in the English Novel: Time, Space, and Narrative* (Princeton: Princeton University Press, 1998).

44. I'm relying on David Burrell's analysis of analogy here that he discusses in *Analogy and Philosophical Language*. See David B. Burrell, *Analogy and Philosophical Language* (New Haven: Yale University Press, 1973).

45. Thomas DiPiero and Timothy J. Reiss both argue that a significant discursive shift in the seventeenth century refocused attention from resemblance to analysis:

> Beginning in the sixteenth century and continuing on toward the end of the seventeenth century, [Reiss] argues, the possibility that discourse could provide an external, objective knowledge of the real structure of the world evolved from within the prevailing discourse of resemblance, which one sought primarily to place the human being in an ordered relationship to the universe not by arranging arbitrary sign, but presenting words as images of things.

I would add to this that the recourse to analogy when analysis fails is another articulation of this discursive shift.

See Thomas DiPiero, "Unreadable Novels: Toward a Theory of Seventeenth-Century Aristocratic Fiction," *Novel: A Forum on Fiction* 38:2–3 (March 2005): 134. See also Timothy J. Reiss, *The Discourse of Modernism* (Ithaca and London: Cornell University Press, 1982).

46. Dipesh Chakrabarty, *Provincializing Europe: Postcolonial Thought and Historical Difference* (Princeton and Oxford: Princeton University Press, 2000), 28.

I. THE ALCHEMY OF EMPIRE

1. I recognize that I am raising the specter of Jürgen Habermas's *Structural Transformation of the Public Sphere* by invoking this phrase, and my reading of Pope is also meant to problematize Habermas's assumptions about the egalitarianism of the periodical press. Anthony Pollock in particular has a gender-inflected critique of this Habermasian fiction that supports my reading of the significance of gendered relations in *The Rape of the Lock*. See Anthony Pollock, *Gender and the Fictions of the Public Sphere, 1690–1755* (New York: Routledge, 2009).

2. Laura Brown, *Ends of Empire: Women and Ideology in Early Eighteenth-Century English Literature* (Ithaca: Cornell University Press, 1993).

3. Alexander Pope, *The Rape of the Lock*, ed. Cynthia Wall (Boston and New York: Bedford, 1998), 6.

4. Suvir Kaul, *Poems of Nation, Anthems of Empire: English Verse in the Long Eighteenth Century* (Charlottesville and London: University Press of Virginia, 2000), 5.

5. Laura Brown, *Alexander Pope* (New York and Oxford: Basil Blackwell, 1985). Ellen Pollak, *The Poetics of Sexual Myth: Gender and Ideology in the Verse of Swift and Pope* (Chicago and London: University of Chicago Press, 1985).

6. Recent works on china usefully complicate the sociopolitical relations between Britain and China. See Eugenia Zuroski Jenkins, *A Taste for China: English Subjectivity and the Prehistory of Orientalism* (New York and Oxford: Oxford University Press, 2013); David Porter, *The Chinese Taste in Eighteenth-Century England* (Cambridge: Cambridge University Press, 2010); and Chi-Ming Yang, *Performing China: Virtue, Commerce, and Orientalism in Eighteenth-Century England, 1660–1760* (Baltimore: Johns Hopkins University Press, 2011).

7. The phrase "China and Peru" invokes the opening lines of Samuel Johnson's *The Vanity of Human Wishes*, and Kaul makes an allusion to this phrase in connection with Johnson's anxiety over the pursuit of empire.

Much earlier, Aphra Behn marshals the same countries in her account of Surinam in *Oroonoko*, claiming that this "continent whose vast extent was never yet known, and may contain more noble earth than all the universe beside; for, they say, it reaches from east to west, one way as far as China, another to Peru." Behn uses the same markers to explain the breadth of this continent, but the function of both China and Peru are not about anxiety but, rather, to give her readers an idea of lost opportunity. She gently admonishes Charles II, noting that had he "but seen and known what a vast and charming world he had been master of in that continent, he would never have parted so easily with it to the Dutch." See Aphra Behn, *Oroonoko*, ed. Janet Todd. (New York and London: Penguin, 2003), 51.

8. Pope, *The Rape of the Lock*, 3.

9. Kaul, *Poems of Nation, Anthems of Empire*, 41.

10. Andre Gunder Frank, *Re-Orient: Global Economy in the Asian Age* (Berkeley and London: University of California Press, 1998), 147–149.

11. Kenneth Pomeranz, *The Great Divergence: China, Europe, and the Making of the Modern World Economy* (Princeton and Oxford: Princeton University Press, 200), 159–160.

12. Kaul, *Poems of Nation, Anthems of Empire*, 26–27.

13. For a particularly nuanced reading of the genre of the "Oriental tale" and *Rasselas's* post-colonial lessons, see Srinivas Aravamudan, *Tropicopolitans:*

Colonialism and Agency, 1688–1804 (Durham and London: Duke University Press, 1999): 202–213.

14. Frank Brady and W.K. Wimsatt, eds., *Samuel Johnson: Selected Poetry and Prose* (Berkeley: University of California Press, 1977), 91.

15. Ibid.

16. Jack Goody, *The East in the West* (Cambridge: Cambridge University Press, 1996), 1–10. Goody goes on to uncover the ways in which syllogism, primarily considered of Greek origin, exists in other epistemologies that predate the Greeks. See "Rationality in Review."

17. Niall Ferguson. *Empire: The Rise and Demise of the British World Order and the Lessons for Global Power* (New York: Basic Books, 2002).

18. See, for example, K.N. Chaudhuri, *Asia before Europe: Economy and Civilisation of the Indian Ocean from the Rise of Islam to 1750* (Cambridge: Cambridge University Press, 1990); Linda Colley, *Captives: The Story of Britain's Pursuit of Empire and How Its Soldiers and Civilians Were Held Captive by the Dream of Global Supremacy, 1600–1850* (New York: Pantheon, 2002); Andre Gunder Frank, *Re-Orient: Global Economy in the Asian Age* (Berkeley: University of California Press, 1998); Richard H. Grove, *Green Imperialism: Colonial Expansion, Tropical Island Edens and the Origins of Environmentalism, 1600–1860* (Cambridge: Cambridge University Press, 1995); and Pomeranz, *The Great Divergence.*

19. Johnson and his contemporaries had good reason to harbor these fears. The Longitude Act of 1714, precipitated by England's greatest naval disaster during the War of Spanish Succession when Sir Cloudesley Shovell's fleet foundered on the treacherous rocks off the Isles of Scilly, had yet to be effectively solved. John Harrison's marine chronometers are largely credited with offering the most accurate reading of longitude, but it wasn't until 1773 that the Royal Society grudgingly acknowledged this and even then, the less accurate lunar models were more popularly used. Successful merchant voyages were largely the result of luck: without accurate knowledge of longitude, one tended to share shipping lanes with all sorts of other traffic, including pirates, privateers, and dangerous competitors.

20. For Thomas Kuhn, this is a process he identifies as "normal science." In fact, the inception of his argument is a call to deconstruct the idea of history as a "repository" of anecdote and think through the complexities that dominant ideology forecloses. Changing the Eurocentric paradigm of global history, one that is structured by Enlightenment epistemology, is the enterprise Frank, Grove, Pomeranz, and others have been working toward for years. In literary studies, and particularly in postcolonial studies, it is crucial to call into question the automatic assumption that Enlightenment thinking

has been the sole province of European endeavor. See Thomas Kuhn, *The Structure of Scientific Revolutions* (Chicago: University of Chicago Press, 1962).

21. See Frank, *Re-Orient*.

22. See Pomeranz, *The Great Divergence*, 156–159.

23. Ibid., 171–172.

24. As Pomeranz points out, Asian merchants competed very successfully with European merchants as long as the Europeans did not use force.

25. I would like to make a distinction between my use of "British" versus "English" in this context. While the seafaring drives of merchants were drawn from Britons inhabiting a pre—and post–Act of Union Britain (1707), the fashions and tastes shaping London culture and social relations, drawn, perhaps, from the various courts of Charles II and Queen Anne, were specifically English.

26. British Library, Add Ms 33979 (ff. 1–10).

27. BL, Add Ms 35262 (ff. 9r–11r).

28. Zaheer Baber, *The Science of Empire: Scientific Knowledge, Civilization, and Colonial Rule in India* (Albany: State University of New York Press, 1996), 41.

29. According to Baber, the references to both the substance *caute* and the surgical method have not been followed up in any detail; however, 'the ancient *Ayurvedic* text, the *Susruta-Samhita* does contain fragmentary references to roughly similar procedures" (80).

30. BL, Add Ms 35262 (ff. 9–11).

31. An anonymously authored article in Gentlemen's Magazine (1794) also referred to Captain Irvine's "Sepoy" (who turned out to be a Maharatta bullock-cart driver named Cowasjee), noting that the Maharatta surgeon was performing a practice that had been in place since "time immemorial." In this description, the "cement" Scott mentioned was explicitly called terra japonica. Later in the early nineteenth century, the anatomist and surgeon Joseph Constantine Carpue published his treatise *An Account of Two Successful Operations for Restoring a Lost Nose from the Integuments of the Forehead . . . Including Descriptions of the Indian and Italian Methods* (London: Longman, 1816), 36–63, in which once again the substance caute is mentioned. Carpue is an interesting figure: often described as an eccentric anatomist, he came to the profession somewhat reluctantly, having considered first the priesthood and later taking over the bookseller business his uncle, Lewis Carpue of Great Russell Street, Covent Garden, schoolfellow and friend of Pope, had established.

32. BL, Add Ms 33979 (ff. 1–10).

33. Ibid.

34. BL, Rare European Manuscripts Collection, Add MS 35262: 14–15.

35. BL, Add Ms 33979 (ff 1–10).

36. BL, REMC, Add Ms 4432: 14–15.

37. I want to add here that crucial historical interventions have been made to demystify the nineteenth-century notion of science and scientific methodology as both universal and objective. In particular, Philip's *Civilizing Natures* is an important study of the relationship between colonialism and science. See Kavita Philip, *Civilizing Natures: Race, Resources, and Modernity in Colonial South India* (New Brunswick, NJ: Rutgers University Press, 2004), 134–139.

38. BL, Oriental and India Office Collections, Nicholas Downton to the East India Company, Nov. 20th, 1614. In *Letters Received by the East India Company from Its Servants in the East*, Vol. II (1613–1614), Document 181: 169.

39. BL, OIOC, Thomas Kerridge to the East India Company, Jan. 20, 1614 (Agamere). In *Letters Received by the East India Company from Its Servants in the East*, Vol. II (1613–1614), Document 235: 298.

40. Linda Colley's latest treatment of British imperialism offers a clear account of the ways in which stories of British captivity dominated much of the putative histories of British imperialism as an unproblematic telos.

41. Chaudhuri, *Asia before Europe*, 99.

42. Colley's account of the spectacular failure of the British in Tangier is a persuasive example of how certain stories get erased or forgotten in the service of the grand narrative (a key point Benedict Anderson made years ago that, obviously, bears repeating). I am arguing for a similar phenomenon occurring in the representation of epistemological exchanges between India and Britain. For example, she writes that the only extant major history of the seventeenth-century British colony of Tangier was written early in the twentieth century by a woman (E.M.G. Routh—working outside the almost exclusively male establishment of historians of empire), and notes that Sir Hugh Chomley, the major engineer of colonial Tangier, has disappeared from *The Dictionary of National Biography*, thus demonstrating "how effectively Britain's sporadic imperial disasters and retreats were expunged from the historical record and from national and even international memory." See Colley, *Captives*, 33.

43. Chaudhuri, *Asia before Europe*, 99.

44. Bruno Latour argues that our use of modernity defines "by contrast, an archaic and stable past . . . it designates a break in the regular passage of time, and it designates a combat in which there are victors and vanquished" (10). Latour understands the ways in which "modernity" invokes, simultaneously, a chronological and achronological state is what I am attempting to

argue here. That is, on the one hand, the idea of modernity forecloses the possibility of a history of science and technology as effectively as the histories attending scientific discoveries (as Thomas Kuhn has identified). And, yet, those histories, despite their foreclosure from current discourse, haven't disappeared but have been simply buried by the dominant ideology. Part of the project I'm engaged in is a way of restoring the multifarious stories that articulate the production of a scientific fact or technological innovation. See Bruno Latour, *We Have Never Been Modern* (Cambridge, MA: Harvard University Press, 1993).

45. Allen Debus's monumental study, *The Chemical Philosophy*, gives the most comprehensive account of the origins of alchemy and development. He claims that the general agreement is that alchemy was a combination of medicine and pharmacology, fueled by a persistent belief in a unified nature, and documented primarily in philosophical works. He argues, however, that alchemy wasn't simply an abstract theoretical paradigm but was rooted in the practical crafts of metal workers and others. Sources for alchemy in antiquity, then, are also in the "surviving texts that illustrate the practical craft traditions of antiquity. In fact, we do find that the oldest surviving works of metal craftsmen combine an emphasis on the change in the appearance of metals with the acceptance of a vitalistic view of nature—a view that included the belief that metals live and grow within the earth in a fashion analogous to the growth of the human fetus. It was to become basic to alchemical thought that the operator might hasten the natural process of metallic growth in his laboratory and thus bring about perfection in far less time than that required by nature." See Allen G. Debus, *The Chemical Philosophy: Paracelsian Science and Medicine in the Sixteenth and Seventeenth Centuries*, Vol. I (New York: Science History Publications, 1977), 4.

46. Pamela Smith, *The Business of Alchemy: Science and Culture in the Holy Roman Empire* (Princeton: Princeton University Press, 1994), 35.

47. Ibid.

48. Mathematics, arguably, also originates from Indian sources; the idea of zero, coming from a Hindu system, was disseminated by Arab merchants to Europe. Brian Rotman writes, for example:

> The mathematical sign we know as zero entered European consciousness with difficulty and incomprehension. It appears to have originated some 1300 years ago in central India as the distinguishing element in the now familiar Hindu system of numerals. From there it was actively transmitted and promulgated by Arab merchants; so that by the tenth century it was in widespread use throughout the Arab Mediterranean. Between the tenth and the thirteenth century the sign stayed within the confines of Arab culture, resisted by Christian Europe,

and dismissed by those whose function it was to handle numbers as an incomprehensible and unnecessary symbol.

Rotman continues to argue that the development of mercantile capitalism in the fourteenth century and the passage of handling numbers from church educated clerks to merchants and artisan-scientists, among others, resulted in the adoption of zero as crucial to trade and technology. Such artisans, were, according to Smith and Debus, ones that were trained in alchemy, the discourse that was partially responsible for deploying mathematics as a more reliable and useful method of accounting. See Brian Rotman, *Signifying Nothing: The Semiotics of Zero* (Stanford, CA: Stanford University Press, 1987), 7.

49. Robert Markley addresses this sublimation in more explicitly economic terms, arguing that the

> ideality of mathematics . . . rests on the privileging of 'pure' mathematics as a general equivalent and on its status as a commodity: mathematical knowledge becomes part of an exchange economy, but this knowledge must present itself as emptied of any inherent value, that is, as a fetish which can be consumed, traded, sold, and commodified.

(Robert Markley, *Fallen Languages: Crises of Representation in Newtonian England, 1660–1740* [Ithaca, NY: Cornell University Press, 1993], 211.)

50. Debus, *The Chemical Philosophy*, 4.

51. Steven Shapin, *A Social History of Truth: Civility and Science in Seventeenth-Century England* (Chicago: University of Chicago Press, 1994), 200.

52. Shapin, *A Social History of Truth*, 203.

53. BL, OIOC, William Edwards to the East India Company by the Hope Dec. 2, 1615 (rec'd) Dec. 20th, 1614, Amadavar. In *Letters Received by the East India Company from its Servants in the East*, Vol. II (1613–1615), Document 177: 151 (emphasis added).

54. BL, OIOC, Nicholas Downton to the East India Company, Nov. 20th, 1614. In *Letters Received by the East India Company from its Servants in the East*, Vol. II (1613–1615), Document 183: 173–4.

55. See Keith Wrightson. *Earthly Necessities: Economic Lives in Early Modern Britain* (New Haven and London: Yale University Press, 2000).

56. Laura Brown's recent reworking of the poem alludes to the figures of liquidity as part of the trope of commercial power. Donna Landry's account of violence in *Windsor-Forest* identifies the poem with the progression of hunting and blood sport:

A precondition for the *Pax Britannica* that is celebrated at the end of the poem would seem to be for the whole colonized world to have been comparably bloodied. Once the British have extended their empire of field sports sufficiently overseas, the colonized will return to the imperial metropolis to gape at the English, who will forever be madly chasing some creature or the other.

Landry's focus is more on the history of sporting and how this affects the invention of the notion of "countryside," and Brown extends the critique of imperial hegemony that she has established in earlier works. Although the two approaches differ, both share an interest in maintaining an understanding of English power in opaque terms: either the blood-soaked hunts that are metonyms of the colonial enterprise, the material commodities scattered throughout the poem in order to mitigate the violence of English colonialism, or even, as Brown argues in *Fables of Modernity*, Pope's understanding of the "flood" (the Thames, the ocean) as a material embodiment of England's global, political and mercantile power. See Laura Brown, *Fables of Modernity: Literature and Culture in the English Eighteenth Century* (Ithaca, NY: Cornell University Press, 2001), and Donna Landry. *The Invention of the Countryside: Hunting, Walking, and Ecology in English Literature, 1671–1831* (New York: Palgrave, 2001), 117.

57. Brown, *Fables of Modernity*, 33.

58. Kaul reminds us of the classical models of satire that invite ethnographic scrutiny as part of their articulation. Thus Pope's arch reference to "our" speech, color, and attire positions New World inhabitants as ridiculous from the outset. However, as Kaul also argues, this reversal of viewpoint indicates a kind of brinksmanship that Pope deploys in *The Rape of the Lock*.

59. Markley, *Fallen Languages*, 133.

2. MORTAR AND THE MAKING OF MADRAS

1. Carolyn A. Barros and Johanna M. Smith, *Life Writings by British Women, 1660–1850: An Anthology* (Boston: Northeastern University Press, 2000). Kindersley's writings also attracted negative attention, especially from Reverend Hodgson—another proof of the seriousness with which readers in England took her letters.

2. This small but important detail may have accounted for Pyke's unsuccessful attempts to manufacture Madrasi mortar back on St. Helena.

3. Carl Nightingale, "Madras, New York, and the Urban and Global Origins of Color Lines, 1690–1750" (unpublished manuscript in possession of the author courtesy of Carl Nightingale).

4. I am making a distinction between "English" and "British" here based on the shift from England to Britain with the second Act of Union of 1707.

5. Nightingale, "Madras, New York, and the Urban and Global Origins of Color Lines," 5.

6. Henry Davison Love, *Vestiges of Old Madras, 1640–1800, Traced from the East India Company's Records Preserved at Fort St. George and the India Office and from Other Sources*, Vol. I (1913; New York: AMS Press 1968), 382.

7. Ibid., 366.

8. British Library, *Philosophical Transactions of the Royal Society*, Vol. 65: 232–233.

9. Love, *Vestiges of Old Madras*, 381.

10. The rules to wear uniforms may well have been to enforce a professional identity, but even so, it is one that is metonymically linked to a national one.

11. It is worth noting that the pieces of eight with which the British East India Company used to pay wages are Spanish in origin. By the seventeenth century this coin had become the international currency, used by Europeans in global trade, which attests to the relative insignificance of British power in India.

12. J. Talboys Wheeler, *History of English Settlements in India* (Calcutta: W. Newman and Co., 1878), 54.

13. Love, *Vestiges of Old Madras*, 424.

14. Nightingale, "Madras, New York, and the Urban and Global Origins of Color Lines," 6.

15. For example, Joyce Chaplin argues that British encounters with indigenous people like the Inuit were a source of constant surprise; incapable themselves of acclimating to the climate, they were astonished that such "backward" non-European cultures had the wherewithal to exist comfortably without the benefit of European techne. See Joyce E. Chaplin, *Subject Matter: Technology, the Body, and Science on the Anglo-American Frontier, 1500–1676* (Cambridge: Harvard University Press, 2001), 15.

16. Love, *Vestiges of Old Madras*, 421.

17. Private trading was anathema to the East India Company primarily because they could not accrue any profit from the fortunes amassed from these traders. The Company had a "legal" (under British law) monopoly on Indian trade; however, powerful administrators found that they could augment their salaries quite considerably if they traded privately with Indian merchants. According to the historian Om Prakesh:

> Among the important private British traders operating from Coromandel during the second half of the seventeenth century were the governors of Madras. Two of these, Elihu Yale and Thomas Pitt, were particularly active and are known to have amassed huge fortunes, estimated in the case of Yale

at a massive £200,000. Other governors with significant private trading interests included Edward Winter, William Langhorn, Streynsham Master, Gulston Addison, Edward Harrison, and Joseph Collet.

See Om Prakash, "British Private Traders in the Indian Ocean in the Seventeenth and Eighteenth Centuries," in Gerald Maclean, ed. *Britain and the Muslim World: Historical Perspectives* (Newcastle upon Tyne: Cambridge Scholars, 2011), 14.

18. Isaac Pyke, "The Method of making the best Mortar at Madras in East India," *Philosophical Transactions*, 231–235.

19. I use the term "sublimate" in order to emphasize the ways in which material transmutates into the abstract: from humble provenance, mortar now becomes a crucial material to demonstrate cultural and racial difference to the British in India.

20. Richard White, "The Nationalization of Nature," *Journal of American History*, 86:3 (December 1999): 3.

21. Miles Ogborn's work on correspondence and the East India Company is especially germane to my argument. He argues that the attempts the English made at dominating India by the power of writing were always upended by their material circumstances. See Miles Ogborn, *Indian Ink: Script and Print in the Making of the English East India Company* (Chicago and London: University of Chicago Press, 2007).

22. White, "The Nationalization of Nature," 2.

23. The October 2007 Yale University Alumni Newsletter featured an article on "Yale's engagement with India." As part of the Incredible India@60 campaign organized by the Confederation of Indian Industry and the Government of India, Yale convened two panels examining the "challenges that India will face in the coming decades and the rise of women leaders in all facets of India and its global diaspora." Both panels were held at the notoriously exclusive Yale Club of New York City, a gesture that solidified Yale President Richard C. Levin's statement that the study of India and South Asia had "blossomed" at Yale during the last decade, and that he expected "expanded exchanges and partnerships with India in the years to come." At the end of the letter was a short column explicating the connections between Yale and India "at a glance," and the first item noted Elihu Yale's tenure as Governor of Madras, and his gift of "books, Indian textiles, and other goods" that resulted in the vaunted institution of the university. See http://www.yale.edu/opa/eline/2007/200710.html.

24. Although most of the diamonds Yale amassed were brokered by Jacques de Paivia, as a Portuguese and as a Jew, de Paivia figured outside the community of British merchants and was, more to the point, a serious rival to their mercantile position.

25. Giles Milton, *Nathaniel's Nutmeg, or, The True and Incredible Adventures of the Spice Trader Who Changed the Course of History* (New York: Penguin, 1999), 3.

26. I am drawing on Alain Badiou's definition of the "event of truth" from *Being and Event* to situate my own biographical supplement of Elihu Yale's life within a theoretical paradigm. Badiou argues that an event is a decision about something undecidable, a "hole' in an established "encyclopedia" of knowledge. Badiou writes: "I call 'encyclopedia the general system of predictive knowledge internal to a situation; i.e. what everyone knows about politics, sexual difference, culture, art, technology, etc..'" Into this text I am putting the genre of biographical representation, one that primarily serves the interests of nation and state, but also of an imperial epistemology. Badiou goes on to identify the expressions that are meant to solidify the uncertain status of floating signifiers (e.g. "illegal immigrants") that are unstable sites and therefore can explode, resulting in what Thomas Kuhn observed a "paradigm shift." Badiou writes: "an event is what decides about a zone of encyclopedic indiscernibility." Elihu Yale's life is a known entity, unquestionably documented in a hard and fast epistemological structure, equally solidified by his metonymical relation to Yale University—another known or "discernible" entity. But putting together the events of other things existing at the margins of knowledge may "explode" the truth claims of hegemonic biographical representations. I am, therefore, supplementing those narratives whose hegemony draws upon nation, state, epistemology, and taxonomy with another look at the status of those ideologies. See Alain Badiou, *Being and Event*, trans. Oliver Feltham (London and New York: Continuum, 2005), 146–147.

27. Steven Shapin and Simon Schaffer raise such an issue in their study on the history of experiments. Focusing on the "great paradigm of experimental procedure," Robert Boyle's interest in pneumatics and air-pumps precisely because of its "canonical character in science texts," their inquiry into the history of these experiments reveals something interesting: "it is entirely appropriate that an excellent account of Boyle's pneumatic experiments of the 1660s constitutes the first celebrated series of *Harvard Case Histories of Experimental Science*," they write, because it provides a "heuristic model of how authentic scientific knowledge should be secured." Because this history now has canonical status, "it has provided a concrete exemplar of how to do research in the discipline, what sorts of historical questions are pertinent to ask, what sorts are not germane, and what general form of historical narrative and explanation ought to be." Their initial questions—What is an experiment? How is an experiment performed? or most importantly "*Why* does one do experiments in order to arrive at scientific truth?"—are foreclosed from

Harvard's history and therefore, as their analysis proceeds, they become "increasingly convinced that the questions we wished to have answered had not been systematically posed by other writers." My point is that the questions posed by Shapin and Schaffer dislodge the canonical history, here institutionalized by Harvard, from its moorings and reveal those moorings to have been produced by the self-image of institutional "history"—knowledge-claims authenticated by the educational status of Harvard. See Steven Shapin and Simon Schaffer, *Leviathan and the Air-Pump: Hobbes, Boyle, and the Experimental Life* (Princeton: Princeton University Press, 1985), 3–4.

28. Milton, *Nathaniel's Nutmeg*, 1–8.

29. See Richard H. Grove, *Green Imperialism: Colonial Expansion, Tropical Island Eden's and the Origins of Environmentalism, 1600–1660* (Cambridge: Cambridge University Press, 1995); E.C. Spary, *Utopia's Garden: French Natural History from Old Regime to Revolution* (Chicago: University of Chicago Press, 2000) and "Of Nutmegs and Botanists: The Colonial Cultivation of Botanical Identity," in *Colonial Botany: Science, Commerce, and Politics in the Early Modern World*, ed. Londa Schiebinger and Claudia Swan (Philadelphia: University of Pennsylvania Press, 2005), 187–203.

Both Grove and Spary are interested in the utopian or Edenic fantasies with which Europeans projected on island cultures that furnished an inexhaustible supply of botanical wealth. While Spary focuses on nutmeg, Grove extends his discussion to the ways in which a number of different plants, replicated on European-owned islands (wrested by colonial force), provided an early model of conservation.

30. Many other historians engage with the spice trade, notably John Keay, C.R. Boxer, Robert Markley, and Anthony Reid, to mention a few. I address the transition of the specific spice nutmeg from commodity to knowledge claim than the politics of the spice trade itself. See John Keay, *The Honourable Company: a History of the English East India Company* (New York: Macmillan, 1991); Charles Ralph Boxer, *From Lisbon to Goa, 1500–1750: Studies in Portuguese Maritime Enterprise* (London: Variorum Reprints, 1984); Robert Markley, *The Far East and the English Imagination, 1600–1730* (Cambridge: Cambridge University Press, 2009) and Anthony Reid, *Southeast Asia in the Age of Commerce, 1450–1680* (New Haven: Yale University Press, 1988).

31. Grove, *Green Imperialism*, 168–171.

32. See Steven Shapin, *A Social History of Truth: Civility and Science in Seventeenth-Century England* (Chicago: University of Chicago Press, 1994). Also Shapin and Schaffer, *Leviathan and the Air-Pump*.

33. E.C. Spary, "Of Nutmegs and Botanists: The Colonial Cultivation of Botanical Identity," in Schiebinger and Swan, *Colonial Botany*, 188.

34. Quoted in ibid., 192.

35. Ibid., 188.

36. I am using the notion of "supplement" here in the Derridean sense—the thing that signifies lack by replacing its initial articulation.

37. This is a soft figure based on Giles Milton's estimates (whose sources are original diaries, journals and letters, mostly collected by Samuel Purchas in *Purchas His Pilgrimes* [1625]). Milton claims:

> In the Banda Islands, ten pounds of nutmeg cost less than one English penny. In London, that same spice sold for more than L2.10s.—a mark-up of a staggering 60,000 per cent. A small sackfull was enough to set a man up for life, buying him a gabled dwelling in Holborn and a servant to attend to his needs." As I will demonstrate, nutmegs were responsible for many private fortunes, in spite of London merchants' concerns about the illegal trade in nutmeg. (*Nathaniel's Nutmeg*, 6)

38. Viswanathan offers a wonderfully lucid and historically complex account of Elihu Yale's various subject positions. My difference from Viswanathan's argument is that I am focusing on the unnamed materials of Yale's bequest and the relation between those abject materials and the process of empire building. See Gaury Viswanathan's "The Naming of Yale College," in *Cultures of United States Imperialism*, ed. Amy Kaplan and Donald E. Pease (Durham and London: Duke University Press, 1993), 85–108.

39. I need to emphasize the fact that the "state" that the Collegiate School occupied was only a small part of the New England colonies—far from being the emerging nation-state that eventually constituted part of the thirteen colonies. Similarly, England's only claim to Madras was the Fort St. George. Both of these emergent states, however, articulate what Richard White identifies as the simultaneous fragility and strength of national histories:

> The dominance of the national in modern histories is both less and more than it seems. Push it a little and its strength appears illusory; push it still more and its strength is all too real. Modern history, it is quite true, has often not literally been national history: it has often been local or regional history. (White, "The Nationalization of Nature," 3)

40. I might add that national histories are themselves the products of a self-congratulatory, Eurocentric writing of sociopolitical modernity.

41. The narratives about Eli Yale's life and legacy sanctioned by the university would necessarily foreclose some of the more dubious aspects of his career, even if the manner of such foreclosure takes shape as the "colorful" exploits of Yale's life.

42. Hiram Bingham is another such "colorful" figure in the history of the university. Credited with being the "discoverer" of Machu Piccu, Bingham capitalized on his position as an anthropologist to provide Yale with his own list of valuable objects. Local "others" who led him to these sites, of course, crucially supplemented his explorations of "unexplored" Incan cities. My point is that the "biography" of Yale University and its founders, donors, and other significant alumnae are very much connected with regional and local histories—of people and objects—that could have surprising connections. As Governor of Connecticut (1922–1924), Hiram Bingham is metonymically connected with Yale's own gubernatorial position. See Hiram Bingham, *Elihu Yale: The American Nabob of Queen Square* (New York: Dodd, Mead & Company, 1939).

43. Love, *Vestiges of Old Madras*, 463–464.

44. Love, *Vestiges of Old Madras*, 491.

45. The economic historian, Om Prakash, has noted that most East India Company factors, governors included, profited from private trade. Although the Directors of the Honourable Company had strict rules against private trading, they were almost impossible to enforce particularly when governors would use their charters as suggestive, not prescriptive, and would shape the rules to suit themselves.

46. J. Talboys Wheeler, *Early Records of British India: A History of the English Settlements in India* (London: Curzon Press, 1878), 84–85.

47. Although it is true that the British East India Company had, on several occasions, ordered a halt to the trade in slaves, it also realized that it was losing money and had decided to profit from it. See Wheeler, *Early Records of British India*, 84–85.

48. Love, *Vestiges of Old Madras*, 502–503.

49. Love, *Vestiges of Old Madras*, 534–545.

50. Bingham, *Elihu Yale*, 296–297.

51. The indignation with which such a breach of cultural propriety was received is fairly understandable in the case of Yale's brother, wife, and children. Yet it seems to have somehow carried through the centuries and is remarkably articulated by Yale's biographer:

> The change seemed to come with the departure of Madame Yale and her three little daughters. As long as she was with Eilhu in India he was respected, modest, faithful and trusted. After she left, he seemed to change. . . . He was domineering, opinionated, aggressive and unable to hold the confidence or the respect of other members of the Council. . . . if the Governor had been Elihu Yale of his early married life, or of his accession to the supreme power as he was before his son died and his wife went back to England, it is not likely that he

would have forced measures through against the opposition of practically all the members of his Council. Something changed him. Did Hieronima de Paivia and her little son have anything to do with it? Possibly. (Bingham, *Elihu Yale*, 273)

52. W.W. Hunter, *History of British India*, Vol. II (New York, Bombay, and Calcutta: Longmans, Green and Co., 1912), 180–181.

53. One of Yale's proposals was to rename the various parts of the Fort:

And first for the Cittadell or Fort which went under the various tearms of Flagstaff Point, Horse Point, Cookroom &tc., and never and Settld certain names Registr'd for them, 'tis now appointed that they goe under the following names for the future, and the Souldiers and Inhabitants be quartr'd to them accordingly, Vzt.: The four Fort Points called *The English Point, Scotch Point, French Point, Irish Point*. (Love, *Vestiges of Old Madras*, 535)

54. Privy Council Registers in the Public Records Office, MSS.

55. The New East India Company and the East India Company eventually merged in 1708.

56. Hiram Bingham, *Elihu Yale: Governor, Collector, Benefactor,* Proceedings: American Antiquarian Society (Worcester, 1937), Vol. 47: 93–144.

57. James C. Scott, *Seeing Like a State: How Certain Schemes to Improve the Human Condition Have Failed* (New Haven: Yale University Press, 1998), 11.

58. Yale University has made conscious decisions to reinforce its distance from Madras: the architectural tribute to Elihu Yale is a replica of the Welsh chapel where he is buried, not the Garden House of White Town where he spent many years amassing the fortune that so crucially benefited the university, in the company of Hieronima Paivia.

59. White, "The Nationalization of Nature," 1.

60. Milton, *Nathaniel's Nutmeg*, 6.

3. ICE AND THE PRODUCTION OF BRITISH CLIMATE

1. George Orwell, *Burmese Days* (Orlando: Harcourt Brace Jovanovich, 1962), 120–121.

2. Robinson Crusoe spends a good deal of time debating the merits of clothing in the tropical climate of his island, finally concluding the necessity for clothing because his skin is not "Mulatto" enough to withstand the sun's rays. Alan Bewell makes a particularly important argument about cholera mapping the lines connecting Britain to its colonial possessions. Citing from the *Lancet*, Bewell notes the several ways in which the biomedical identity of cholera became a permanent feature of the Indian landscape in the English imaginary. See Alan Bewell, *Romanticism and Colonial Disease* (Baltimore:

Johns Hopkins University Press, 1999), 242–276. For another illuminating account of this complicated node of climate, sexuality, and imperial identity, see Felicity A. Nussbaum, *Torrid Zones: Maternity, Sexuality, and Empire in Eighteenth-Century English Narratives* (Baltimore: Johns Hopkins University Press), 1995.

3. Orwell, *Burmese Days*, 65.

4. Emmanuel Le Roy Ladurie, *Times of Feast, Times of Famine: a History of Climate Since the Year 1300*, trans. Barbara Bray (London: George Allen & Unwin, 1971), 138–140.

5. Sir Robert Barker, "The Process of Making Ice in the East Indies," *Philosophical Transactions* XXII (1775): 257.

6. *Philosophical Transactions of the Royal Society* 65: 256.

7. Ladurie, *Times of Feast, Times of Famine*, 138.

8. Brian Fagan, *The Little Ice Age: How Climate Made History, 1300–1850* (New York: Basic Books, 2000), 88.

9. Ibid., 91.

10. Steven Shapin and Simon Schaffer, *Leviathan and the Air-Pump: Hobbes, Boyle, and the Experimental Life* (Princeton: Princeton University Press, 1985), 15.

11. For example, Parlington Hall, seat of the Baronet Sir Edward Gascoigne (1743), was famous for architectural innovations including the ice house in which ice from nearby ponds and lakes was stored, and could keep up to three years. Initial conjectures suggest that the ice was used to keep sides of venison and beef, but in addition to this form of refrigeration, the ice stored in the house was also used to cool white wines and other drinks for summer consumption. A contemporary of Barker's, Gascoigne's experiment with ice storage would very likely have been in Barker's mind during his time in India.

12. Barker, "The Process of Making Ice in the East Indies," 252–257.

13. Joyce E. Chaplin, *Subject Matter: Technology, the Body, and Science on the Anglo-American Frontier, 1500–1676* (Cambridge, MA: Harvard University Press, 2001), 15.

14. "The English even cheered themselves by using the 'Little Ice Age' to emphasize similarity between the new world and the old. The ice that one northbound crew encountered in 1607 was explainable, they said, because that was the year 'when extraordinary frost was felt in most parts of Europe" (ibid., 50).

15. Ibid., 51.

16. See Linda Colley, *Captives: Britain, Empire, and the World, 1600–1850* (New York: Pantheon, 2002).

17. This connection is not simply allusive. In a study considering the effects of the Asian southwest monsoon on the North Atlantic Ocean, Anil K. Gupta and colleagues claim:

> During the last ice age, the Indian Ocean southwest monsoon exhibited abrupt changes that were closely correlated with millennial-scale climate events in the North Atlantic region, suggesting a mechanistic link. In the Holocene epoch, which had a more stable climate, the amplitude of abrupt changes in North Atlantic climate was much smaller, and it has been unclear whether these changes are related to monsoon variability. Here we present a continuous record of centennial-scale monsoon variability throughout the Holocene from rapidly accumulating and minimally bioturbated sediments in the anoxic Arabian Sea. Our monsoon proxy record reveals several intervals of weak summer monsoon that coincide with cold periods documented in the North Atlantic region4—including the most recent climate changes from the Medieval Warm Period to the Little Ice Age and then to the present. We therefore suggest that the link between North Atlantic climate and the Asian monsoon is a persistent aspect of global climate.

> Although India did not experience the general cooling that northern Europe and the northeastern coast of the Americas did, nevertheless the "several intervals of weak summer monsoons" coincided with the climate changes defining the Little Ice Age.

See Anil K. Gupta, David M. Anderson, and Jonathan T. Overpeck, "Abrupt Changes in the Asian Southwest Monsoon during the Holocene and Their Links to the North Atlantic Ocean," *Nature* 421 (January 23, 2002): 354–357. For more on the scientific work on global climatology, and the Asian monsoons in particular, see R. Bin Wang, *The Asian Monsoon* (Chichester, UK: Praxis, 2006).

18. I want to be clear here that I'm not using the Little Ice Age to collapse human perspectives from the sixteenth to the nineteenth centuries; rather, I'm arguing that climate conditions these perspectives.

19. Frank Brady and William K. Wimsatt, eds., *Samuel Johnson: Selected Poetry and Prose* (Berkeley: University of California Press, 1977), 291.

20. Ibid., 277.

21. Rudyard Kipling's poem "The White Man's Burden" is a pithy example of imperial drudgery.

22. Brady and Wimsatt, *Samuel Johnson*, 277–278.

23. Rajani Sudan, "Lost in Lexicography: Legitimating Cultural Identity in Johnson's 'Preface' to the 'Dictionary,'" *ECTI* 39:2 (Summer 1998): 127–146.

24. Brady and Wimsatt, *Samuel Johnson*, 296.

25. Ellis's admonition of the butler in *Burmese Days* may not be simply rhetorical. I've argued for the material connection between Johnson's lexicography and the Little Ice Age. Orwell's novel, written some eighty years after its end, juxtaposes melting ice and the butler's fluency in English. Swallowing dictionaries, as Ellis's picturesquely figures it, comes about when British imperial authority has become too lenient, its borders unpatrolled and unguarded. Thus Johnson's fear that "commerce . . . depraves the manners, corrupts the language" is enacted in this scene.

26. Barker, The Process of Making Ice in the East Indies," 252–257.

27. Shapin and Schaffer, *Leviathan and the Air-Pump*, 21.

28. Brady and Wimsatt, *Samuel Johnson*, 296.

29. Ibid., 298.

30. Ibid., 294–295.

31. Kavita Philip, *Civilizing Natures: Race, Resources, and Modernity in Colonial South India*, (New Brunswick, NJ: Rutgers University Press, 2004), 35.

32. Ibid., 40.

33. I use tabula rasa to refer to Barker's fantasies of the profit to be accrued by the manufacture of ice through Indian techne. Of course, the British were extraordinarily sensitive to the deleterious effects South Asian climate wrought upon their health.

34. Philip, *Civilizing Natures*, 40.

35. Scott's compelling example invokes the environmental history of mono-cropping trees in medieval Germany, the ways in which abstract administrative decisions, made without accounting for local issues, changed the history of the land. See James C. Scott, *Seeing Like a State: How Certain Schemes to Improve the Human Condition Have Failed* (New Haven: Yale University Press, 1998).

36. Originally under the jurisdiction of the kingdoms of Nepal and Sikkim, Darjeeling became folded into West Bengal when member of the East India Company, traveling to Sikkim in 1828, decided that the area was an admirable site for a sanitarium for British soldiers. Darjeeling grew in popularity with the British ruling class and Maharajahs of local princely estates. The legacy, however, of this administrative decision, is one that is still being felt; British appropriation of topography regardless of ethnic and political divisions, affected and continues to affect the political climate of West Bengal.

37. Henry David Thoreau, *Walden*, ed. Jeffrey S. Cramer (New Haven and London: Yale University Press, 2004), 287–288.

38. Henry David Thoreau, *Journal*, Vol. 2: *1842–1848*, ed. Robert Sattel-meyer (Princeton: Princeton University Press, 1984), 371.

39. Orestes A. Brownson, "The Everlasting Yes," in *The Transcendentalists: An Anthology*, ed. Perry Miller (Cambridge, MA and London: Harvard University Press, 1977), 46–47.

40. Ibid., 46.

41. Henry D. Thoreau, "The Natural History of Massachusetts," in Miller, *The Transcendentalists*, 325.

42. Born into a family whose fabled wealth derived from Jamaican slave plantations, Eric Arthur Blair's experience with British colonial culture was complex. He was born in colonial India of a father who worked in the opium department of the Indian Civil Service and a mother who was raised in Burma. After being educated in England, Blair returned to Burma in 1922 to serve in the Indian Imperial Police for the next six years until the dengue fever he contracted cut his career short. Despite his fame as a republican supporter, it seems more than likely that Blair was integrated into the ideologies of colonial hegemony, something that he spent the rest of his career as a journalist resisting, perhaps under cover of his nom de plume, George Orwell.

43. Orwell, *Burmese Days*, 24.

44. Julia Kristeva, *Powers of Horror: An Essay on Abjection*, trans. Leon S. Roudiez (New York: Columbia University Press, 1982), 2.

45. Orwell, *Burmese Days*, 34.

46. Fagan, *The Little Ice Age*, 91.

47. By "reconfigured," I am alluding to the shift from early eighteenth-century British East India Company factors who had a precarious purchase on the Coromandel Coast, to the later eighteenth-century East India Company whose triumph over Tipu, Sultan of Mysore, ushered in an new age of British colonial power.

4. INOCULATION AND THE LIMITS OF BRITISH IMPERIALISM

1. Alan Bewell, *Romanticism and Colonial Disease* (Baltimore and London: Johns Hopkins University Press, 1999), 131–160.

2. Sally Shuttleworth has been particularly prolific in this area. Her contribution is an extraordinary representation of the parallel developments of medicine as a science and the dissemination of quackery. See her "Female Circulation: Medical Discourse and Popular Advertising in the Mid-Victorian Era," in *Body Politics: Women and the Discourses of Science*, ed. Mary Jacobus, Evelyn Fox Keller, Sally Shuttleworth, (New York and London: Routledge), 47–66.

3. See Andrea Rusnock, *Vital Accounts: Quantifying Health and Population in Eighteenth-Century England and France* (Cambridge: Cambridge University Press, 2002), 44–45.

4. Lady Mary Wortley Montagu, *Embassy to Constantinople: The Travels of Lady Mary Wortley Montagu*, ed. Christopher Pick (New York: New Amsterdam Books, 1988), 121.

5. Ibid., 121.

6. Lady Mary Wortley Montagu's intentions to inoculate her son are profoundly overdetermined. Her relationship with Edward Wortley Montagu Jr. has been interestingly and complexly discussed by Bernadette Andrea, who "slandered him as a 'rogue, a 'rake,' and an 'animal,' considering him a general disappointment." Arguing that "Edward remained a little more than a cipher in this episode [of his inoculation]," Andrea claims that "he later used it to his advantage when he claimed the operation 'infused something of Turkish blood into my English veins.' He thereby evoked the still current belief that the mingling of blood determined one's identity to authenticate his return to the Ottoman Empire fifty years later as a 'Turk'—a synonym for Muslim throughout the early modern period." Andrea's argument establishes the self-consciousness with which these early modern travelers embraced visceral notions of alterity, something that shifted quite radically in nineteenth-century ideologies of nation. See Bernadette Andrea, "Alternatives to Orientalism? Mary Wortley Montagu and Her 'Turkish' Son," in *Britain and the Muslim World: Historical Perspectives*, ed. Gerald Maclean (Newcastle upon Tyne: Cambridge Scholars Publishing, 2011), 119–120.

7. Montagu, *Embassy to Constantinople*, 121.

8. Rusnock, *Vital Accounts*, 45.

9. Montagu, *Embassy to Constantinople*, 121–122.

10. Denys Van Renen argues that Lady Mary Wortley Montagu "challenges a world order that seeks to carve the world up into European-controlled territories" in part by using her aristocratic position to buttress her authority. See Denys Van Renen, "Montagu's Letters from the Levant: Contesting the Borders of European Selfhood," *Journal for Early Modern Cultural Studies* 11:2 (Fall 2011): 1–34.

11. Srinivas Aravamudan, *Tropicopolitans: Colonialism and Agency, 1688–1804* (Durham and London: Duke University Press, 1999), 184–185.

12. Fernand Braudel, *The Mediterranean and the Mediterranean World in the Age of Philip II* (New York: Harper and Row, 1979).

13. Bernadette Andrea has done extensive work on the role of Edward Wortley Montagu, Jr. as an alternative (and corrective) to the paradigm of orientalism Edward Said introduced.

14. Daniel Defoe, *A Journal of the Plague Year*, ed. Louis Landa (Oxford: Oxford University Press, 2010), 3.

15. Ibid.

16. The Royal College of Physicians only accepted graduates of Oxford and Cambridge into their elite society, thus guaranteeing that the practice of medicine was the province of the nobility. This didn't mean that others couldn't practice medicine—the field was unregulated by any governmental process—but that the formal appeal Lady Mary Wortley Montagu made was to an audience composed of her own class.

17. "You will be surprised," she writes to Anne Thistlethwayte in April of 1717, "at an account [of Turkish homes] so different from what you have been entertained with by common voyage-writers who are fond of speaking of what they don't know." Montagu, *Embassy to Constantinople*, p. 125.

18. In addition to Aravamudan, see Bernadette Andrea, *Women and Islam in Early Modern English Literature* (Cambridge: Cambridge University Press, 2007).

19. British Library, European Manuscripts, Add Ms 4432 271–272.

20. Zaheer Baber, *The Science of Empire: Scientific Knowledge, Civilization, and Colonial Rule in India* (Albany: State University of New York Press, 1996), 41.

21. Ibid., 41–42.

22. Ibid., 8.

23. BL, EM, Add Ms 4432 285–288.

24. I am well aware of the fact that there are similar claims that locate the origins of inoculation in parts of Africa and China. Clearly, techne is not the province of a single territory and simultaneous discoveries are legion. My interests, however, are in the Vedic origins of this practice. See Eugenia W. Herbert, "Smallpox Inoculation in Africa," *Journal of African History* 16:4 (1975): 539–559.

25. The Vaidhya Bhramins were the group of outcast Brahmins who continued their medical practice. They organized themselves as sects of *caranavaidhya* or "roving physicians." See Baber, *The Science of Empire*, 41.

26. BL, EM, Add Ms 4432 285–288.

27. Ibid.

28. One way of accounting for the Turkish origin of smallpox inoculation is to imagine the transmission of techne via trade routes, the same way that "Arabic" numerals (including the concept of zero) moved from India to Asia Minor via Arabic merchants.

29. What is remarkable is the persistence with which the disavowal of India's medical contribution to the history of smallpox inoculation continues. Putatively writing a comprehensive history of "healers," Kurt Pollak, for example, makes a passing reference to the Indian origins of smallpox inoculation without the slightest elaboration: "Preventive inoculation against

smallpox, which was practised in China from the eleventh century, apparently came from India." Granted this history was written in 1963, but even in the section headed "Doctors in India" there is no mention of the Vaidhyan practice. See Kurt Pollak, *The Healers: The Doctor, Then and Now* (London: Thomas Nelson and Sons, 1963), 37.

30. Nadja Durbach, *Bodily Matters: The Anti-Vaccination Movement in England, 1853–1907* (Durham and London: Duke University Press, 2005), 21.

31. BL, EM, Add Ms 4432 285–288.

32. For a fuller history of Holwell's long residence in India and the practice of variolation in India, see David Arnold, *Colonizing the Body: State Medicine and Epidemic Disease in Nineteenth-Century India* (Berkeley and London: University of California Press, 1993), 127–133.

33. In the Royal Society Papers, for example, Cromwell Mortimer discusses the distemper "raging among the Cow-kind in the Neighbourhood of London," noting in particular that

> Dr. Lobb (?), a very diligent & laborious observer, of what occurs in his profession, as his histories of various cases of the smallpox & his curious Experiments on Dissolvents of the Stone sufficiently evince, has lately published in a Collection of Letters relating to the plague, an Enquiry into the Quality of the cause of the Contagious Sickness among the Cattle: it were to be wisht what he proposes were now tried.

This letter, dated November 21, 1745, suggests that the connection between smallpox and cowpox was, in fact, being made by people before Jenner's discoveries. BL, Add MS 4438.

34. Smallpox inoculation was also practiced in China and Africa, although the methods varied. Most notably, George Washington ordered his army to be inoculated against smallpox; he may well have observed the African method when he was in Barbados.

35. I want to add that, given its Brahmin origins and the sacred status of the cow in Hindu culture, it is a deep irony that cowpox replaced smallpox as the matter of inoculation.

36. See Andrea Rusnock, "Catching Cowpox: The Early Spread of Smallpox Vaccinations, 1798–1810," *Bulletin of the History of Medicine* 83 (2009): 17–36.

37. William Tebb, "Sanitation, Not Vaccination: The True Protection against Small-Pox," paper presented at the Second International Congress of Anti-Vaccinators, Cologne, October 12, 1881.

38. *National Anti-Compulsory Vaccination Reporter,* October 1, 1883.

39. Durbach, *Bodily Matters,* 79.

40. Bewell, *Romanticism and Colonial Disease,* 244.

41. Durbach, *Bodily Matters* 153.

42. Martin quoted in Bewell, *Romanticism and Colonial Disease*, 245.

43. Durbach, *Bodily Matters* 153.

44. Mary Hume-Rothery, John Pickerton who edited the *Anti-Vaccinator and Public Health Journal*, and J.J. Garth Wilkinson, all virulent anti-vaccinationists, posed these arguments.

45. Charlotte Brontë, *Jane Eyre*, ed. Richard J. Dunne (New York and London: Norton, 1993), 301.

46. See Rajani Sudan, *Fair Exotics: Xenophobic Subjects in English Literature, 1720–1850* (Philadelphia: University of Pennsylvania Press, 2002).

47. John Cleland, *Fanny Hill, or Memoirs of a Woman of Pleasure*, ed. Peter Wagner (New York: Penguin, 1985), 40.

48. Cleland's insistence on Fanny's virtue is an ironic response to Richardson's spotless heroines; however as is the case for most satire, fondness also underwrites its points (emphasis added).

49. William H. Epstein, *John Cleland: Images of a Life* (New York: Columbia University Press, 1974), 57.

50. Ibid., 23.

51. Hal Gladfelder, *Fanny Hill in Bombay: The Making and Unmaking of John Cleland* (Baltimore: Johns Hopkins University Press, 2012), 5. I should also note that that there was a good deal of controversy surrounding the dangers of smallpox. John Arbuthnot's 1733 account, *An Essay Concerning the Effects of Air on Human Bodies*, explicitly states that perils of smallpox pales in comparison to malaria and yellow fever.

52. Epstein, *John Cleland*, 32–33.

53. Robert Markley examines the firsthand accounts that John Fryer, John Ovington, and Alexander Hamilton wrote describing their experiences in Bombay, an island (or series of islands) they represent as saturated with environments of disease. See Robert Markley, "'A Putridness in the Air': Monsoons and Mortality in Seventeenth-Century Bombay," *Journal for Early Modern Cultural Studies* 10:2 (Fall/Winter 2010): 106.

54. Ibid.

55. Gladfelder, *Fanny Hill in Bombay*, 18.

56. See Jayant Banthia and Tim Dyson, "Smallpox in Nineteenth-Century India," *Population and Development Review* 25:4 (December 1999): 649–680.

57. Cleland told James Boswell in 1772 that he had composed the novel in Bombay; there is no conclusive evidence that this was in fact so, but what is clear is the proximity between his life in India and the narrative written in the Fleet Street prison.

58. Gladfelder, *Fanny Hill in Bombay*, 39.

59. Ibid.

60. Shuttleworth, "Female Circulation p. 56.

61. Bewell, *Romanticism and Colonial* Disease, 244.

62. The borders of England's cultural imperialism had to be policed as carefully as its colonial ones: harbored in the heart of the Scottish Highlands is the possibility of a Stuart insurrection.

63. Bram Stoker, *Dracula*, eds. Nina Auerbach and David Skal (New York: Norton, 1997), 25.

64. "Bradshaw" is short for *Bradshaw's Guide*, a series of railway timetables developed by the printer (and publisher and cartogrpher) George Bradshaw in the late 1830s.

65. The manufacture of *wootz* is recorded by Helenus Scott in the *Philosophical Transactions* about the same time as John Zephaniah Holwell's treatise on smallpox inoculation. Scott claimed that *wootz* "appears to admit of a harder temper than anything we are acquainted with." See *Philosophical Transactions*, Vol. 85. In fact *wootz* was of great interest to many people, including Benjamin Heyne, the Moravian-Scots surgeon, naturalist, and botanist, who wrote that *wootz* "promises to be of importance to the manufacturers of Britain."

See Benjamin Heyne, *Tracts, historical and statistical, on India: with journals of several tours through various parts of the peninsula: also, an account of Sumatra, in a series of letters* (London: Robert Baldwin, and Black, Perry, and Co., 1814), 363.

Dharampal, the renowned historian of Indian science, writes:
  Till well into the nineteenth century Britain produced very little of the steel it required and imported mostly from Sweden, Russia, etc. . . . Possibly such a lag also resulted from Britain's backwardness in the comprehension of processes and theories on which the production of good steel depended. Whatever may have been the understanding in other European countries regarding the details of the processes employed in the manufacture of Indian steel, the British, at the time *wootz* was examined and analysed by them, concluded "that it is made directly from the ore . . ." . . . . It was only some three decades later that this view was revised. An earlier revision in fact, even when confronted with contrary evidence as was made available by other observers of the Indian techniques and processes, was an intellectual impossibility. "That iron could be converted into cast steel by fusing it in a close vessel in contact with carbon" was yet to be discovered, and it was only in 1825 that a British manufacturer "took out a patent for converting iron into steel by exposing it to the action of carboretted hydrogen gas in a close vessel, at very high temperature."

See Dharampal, *The Beautiful Tree: Indigenous Indian Education in the Eighteenth Century*, Vol. III (Goa: Other India Press, 2000), xix.

66. Stoker, *Dracula*, 326.
67. Ibid., 258–259.
68. Ibid., 258.
69. Ibid., 100.
70. Shuttleworth, "Female Circulation p. 54.
71. John Forbes, ed., *Cyclopaedia of Practical Medicine*, Vol. 3 (London: Whittaker, Treacher & Co., 1834), 745.
72. Scheele was an accomplished organic chemist although he was an apothecary by profession and one of his most famous discoveries was what he called "fire-air," or oxygen. Obviously the provenance of this element is multiple and constitutes a crucial example of Kuhn's paradigm shifts in the process of normal science.
73. Cream of tartar increases the volume of egg whites and cream as well as stabilizing them; it also prevents sugars from crystallizing and is therefore used in syrups and frostings, and mixed with vinegar and water into a paste it becomes a cleaning solution for metals. These uses were well known by eighteenth- and nineteenth-century Britons.
74. Bewell, *Romanticism and Colonial Disease*, 4.
75. See Sander Gilman, *Disease and Representation: Stereotypes of Sexuality, Race, and Madness* (Ithaca: Cornell University Press, 1985).

5. "PLAISTERS," PAPER, AND THE LABOR OF LETTERS

1. *The New Family Receipt Book* (London, for John Murray, 1810).
2. Most of Murdoch's inventions were related to the steam engine and various forms of machinery, although he is probably most famous for his application of gas lighting that replaced oil lamps and candles. His collaborations with James Watt, which almost certainly occurred because of Murdoch's employment as the senior machinist for Boulton and Watt steam engines, are strangely undocumented by correspondence between the two, which raises man questions about intellectual property, particularly since Watt was infamously jealous of his own patents.
3. *Philosophical Transactions*, 99–100.
4. In 1754, the East India Company recorded this shortage in the following treatise:

This deprivation did in fact take place during the last war and great inconvenience was sustained by British Shipping, and the price of Hemp which in 1792, was only £25 per ton, rose to £118 in 1808, an only 259, 689 cwt (?) were imported in that Year. The Colonies therefore and India were looked to for a supply of Hemp and its cultivation encouraged in North America.

The Cultivation of Hemp in India obtained very great attention from the Court of Directors and instructions were sent to the Governments there to encourage the growth, as well as that of other Cordage Plants. As the Natives of India employ between 40 and 50 different kinds of Plants for the fibre which they yield fitted for this purpose in different degrees; the subject of investigation was sufficiently extensive, and received great ["great" added here] attention from Dr. Roxburgh. A few only however of the Cordage plants of India are extensively cultivated in that Country or known in commerce, as *Coir, Sun Sunnee* or Brown Indian Hemp & *Jute*. . . . On the present occasion I confine myself to the Hemp Plant itself (the *Cannabis Sativa* of Botanist-?-) as being the most valuable of the whole, and because it is in general erroneously supposed that it can only be successfully cultivated in European regions, though there is every reason to believe that it is originally a Native of Asia—and even that its Greek and Latin name *Cannibis* is derived from the Arabic *Kinnub.*

See BL, India Office Collection F/4/1754/71645.

5. Claudia L. Johnson, *Jane Austen: Women, Politics, and the Novel* (Chicago and London: University of Chicago Press, 1988), 121.

6. Jane Austen, *Emma*, eds. Richard Cronin and Dorothy McMillan (Cambridge: Cambridge University Press, 2005), 366. Subsequent references to Emma will be made from this edition and will be noted in parentheses.

7. Nicholas Dames reads the destruction of Harriet's relics as a necessary destruction of common property between Emma and Harriet that brings nostalgia "into contact with the more drastic excisions of absolute forgetting." I am persuaded by this reading; my interest, however, is with the conditions these relics embody—not simply as metonymies—before they are consigned to the firs. See Nicholas Dames, *Amnesiac Selves: Nostalgia, Forgetting, and British Fiction, 1810–1870* (Oxford: Oxford University Press, 2001), 40–43.

8. J.A. Downie has written on the attempts to "re-position [Austen's] novels within a bourgeois or 'middle—class framework,'" arguing that Austen's preoccupation with issues of social position reflect the family's own association with the aristocracy as part of the gentry which were historically associated, in Britain, with nobility. See J.A. Downie, "Who Says She's a Bourgeois Writer?: Reconsidering the Social and Political Contexts of Jane Austen's Novels," *Eighteenth-Century Studies* 40:1 (Fall 2006): 79.

9. Edward Copeland's work on Austen and money, of course, provides another context for reading the importance of commodity culture in Jane Austen's work. Specific to *Emma*, however, is his contention that the "very goods—novels among the first—that set themselves up as signs of social truths." I extend his contention by arguing, in the context of *Emma*, *writing* is the form of labor that is disavowed in the production of novel-as-commodity

(one of the "goods"), just as the labor of writing letters that Lieutenant Colonel Ironside engages in his treatise on the Indian sun-plant is disavowed in the production of a scientific treatise published in the Royal Society's *Philosophical Transactions*. See Edward Copeland, "Money," in *The Cambridge Companion to Jane Austen*, ed. Edward Copeland and Juliet McMaster (Cambridge: Cambridge University Press, 2010; online edition 2011), http://dx.doi.org/10.1017/CCO9780521763080, 138. See also Edward Copeland, *Women Writing about Money: Women's Fiction in England, 1790–1820* (Cambridge and New York: Cambridge University Press), 1995.

10. Peter Knox-Shaw also notes the historical timing of this novel commenting on the prospect of peace following Napoleon's abdication in 1814—however illusory—may have prompted Austen to "return to the pacific settings of the earlier fiction." See Peter Knox-Shaw, *Jane Austen and the Enlightenment* (Cambridge: Cambridge University Press, 2004), 197.

11. There are other critics of Austen whose evaluations and commentary are far more sophisticated than Goldwin Smith's; but I am interested in his claim that Austen's "meanings" are so transparently available to readers, nothing remains "hidden" as I will go on to demonstrate. There is another, more sentimental, reason for my use of Smith's work: Goldwin Smith Hall houses both the Department of English at Cornell University where I did my doctoral work and one of the most useful salons of my life: a coffee-house called the Temple of Zeus where I first made Austen's acquaintance.

12. Here I must note Betty A. Schellenberg's work that extends Margaret J.M. Ezell's important study, *Writing Women's Literary History* and Paula McDowell's *The Women of Grub Street: Press, Politics, and Gender in the Literary Marketplace, 1678–1730*. Schellenberg "come[s] to believe that the problem now lies, not so much in a lack of evidence about these women's professional lives, but rather in our continued attempt to fit the evidence into habitual frames of reference." To clarify my own position, I'm not as interested in the *fact* of Austen's literary modesty as I am in its ideological entrenchment; it is to this end that I conjure the illusion of her retiring representation. See Betty A. Schellenberg, *The Professionalization of Women Writers in Eighteenth-Century Britain*, (Cambridge and New York: Cambridge University Press, 2005).

13. Virginia Woolf, *A Room of One's Own* (New York: Harcourt Brace Jovanovich, 1929), 70.

14. Sandra Gilbert and Susan Gubar, *The Madwoman in the Attic* (New Haven and London: Yale University Press, 1979), 154–155.

15. Gilbert and Gubar argue that Austen's cover story of "the necessity for silence and submission" corresponded with the legal condition of married

women whose legal existence is suspended, "covered" by her husband's name and identity.

16. Gilbert and Gubar, *The Madwoman in the Attic*, 154.

17. Margaret Ezell, however, uncovers Virginia Woolf's historiographical limits, arguing that the social histories produced since *A Room of One's Own* reveal that women were neither married off at an early age and therefore silenced nor were they averse to bringing themselves forward into print. "We," she writes, "on the other hand, have taken a text designed to be provocative and to stimulate further research into women's lives in the past and canonized it as history." Included in the "we" are literary critic Sandra Gilbert and Susan Gubar who have, among others, depended on Whiggish histories for their interpretation. My use of these critics is not to establish historical certainties. Rather, I find the insistent representation of women's labor as acts of bondage in Austen, Woolf, Gilbert, and Gubar provocative. See Margaret J. M. Ezell, *Writing Women's Literary History* (Baltimore: Johns Hopkins University Press, 1993), 50.

18. Jan Fergus notes in her fine biography that Austen's life as an author is not the furtive and anxious one Woolf represents. Fergus also notes the amount of work involved with producing a book; focusing on the material conditions of authorship, she writes:

> Considerable drudgery was necessary. As Burney and Hawkins knew too well, authors had to write out fair copies of their works for their publishers entirely by hand, using a quill pen, handmade paper, and ink that they may have mixed for themselves from cakes purchased at a stationer's shop. Furthermore, Austen's lifetime occurred at the very end of the hand press period, when every stage of book production was performed by hand. Although powered printing machinery was being used in England before Austen's death in 1817, all her novels were printed in the traditional way, on hand presses similar in principle to those used by Gutenberg more than 300 years earlier. (Jan Fergus, *Jane Austen: A Literary Life* [Great Britain: Macmillan, 1991], 19)

19. It is also well worth noting that those past nine years coincide with the years since the slave trade was abolished by the 1807 Parliamentary Act, if we presume the novel's "present" to be the year that Austen began writing the novel (1814) and its publication (1816). Female labor—what Jane Fairfax will later term "governess trade"—is a metonym for slave labor, an association that becomes clearer later in the novel.

20. Much has been written on mothering in nineteenth-century India and the crucial place the ayah occupied. For further discussion, see Felicity A. Nussbaum, *Torrid Zones: Maternity, Sexuality, and Empire in Eighteenth-Century English Narratives* (Baltimore: Johns Hopkins University Press, 1995);

Sara Suleri, *The Rhetoric of English India* (Chicago and London: University of Chicago Press, 1992); Jenny Sharpe, *Allegories of Empire: The Figure of the Woman in the Colonial Text* (Minneapolis: University of Minnesota Press, 1993); Joyce Grossman, "Ayahs, Dhayes, and Bearers: Mary Sherwood's Indian Experience and Constructions of Subordinated Others," *South Atlantic Review* 66:2 (Spring 2001): 14–44.

21. Fernand Braudel, *The Structures of Everyday Life.* Vol. I (New York: Harper and Row, 1979), 28.

22. As William Galperin has noted, Austen is not entirely secluded from the political ramifications of the Napoleonic Wars; her brother Frank served as a naval officer under Nelson's command as well as a captain of the East India Company. Her relationship to hemp is thus even more compelling.

23. James Thompson, *Models of Value: Eighteenth-Century Political Economy and the Novel* (Durham and London: Duke University Press, 1996), 185–186.

24. For other compelling accounts of Austen's politics, see Edward Neill, *The Politics of Jane Austen* (London and New York: Macmillan, 1999) and David Monaghan, ed. *Jane Austen in a Social Context* (Totowa, NJ: Barnes and Noble Books, 1981).

25. Janine Barchas's study of "matters of fact" in Jane Austen's corpus addresses the historical specificities of Austen's fiction. She points out that Austen embraced celebrity associations, and argues that the eighteenth century "brought us the 'it' factor." I expand on the notion of "it" in this chapter, arguing that this factor can also be usefully read as an object. See Janine Barchas, *Matters of Fact in Jane Austen: History, Location, and Celebrity* (Baltimore: Johns Hopkins University Press, 2012), 5.

26. William H. Galperin, *The Historical Austen* (Philadelphia: University of Pennsylvania Press, 2003),183. Galperin astutely remarks on Miss Taylor's agency in the prehistory of her marriage, suggesting that Emma's role in this affair is largely self-congratulatory, while Miss Taylor's "exit strategy from quasi-servitude" seems the more insistent spur. My only addition to this point is to suggest that her servitude is not quite "quasi" but actual, if we equate, as the novel does, the number of spaces within the home as to the amount of her obligation.

27. Nancy Armstrong, *Desire and Domestic Fiction: A Political History of the Novel* (New York: Oxford University Press, 1987), 148.

28. Galperin, *The Historical Austen*, 182.

29. A compelling parallel exists between the laws of matrimony, which include courtship, the law of coverture, and state sovereignty. "Living within the state," writes James Scott, "meant, virtually, by definition, taxes, conscription, corvee labor, and, for most, a condition of servitude." As members of the

cultural periphery, Miss Taylor, Harriet Smith, and Jane Fairfax must be folded into the imperial metropole, which is most efficiently enacted through courtship. The "truly imperial project" Scott identifies—transforming the "periphery" into a "fully governed, fiscally fertile zone" is made possible "only by distance-demolishing technologies." The examples Scott cites are fairly recognizable forms of colonial labor—"all-weather roads, bridges, railroads"—but what is not included in a list that expands to GPS systems is letter writing: the technology most available to women. See James C. Scott, *The Art of Not Being Governed: An Anarchist History of Upland Southeast Asia* (New Haven and London: Yale University Press, 2009), 7–11.

30. Although it seems as if Harriet Smith's courtship is as distant as possible from Emma's—the former part of the yeomanry, the latter now part of Knightley's seat—one needs to remember that Emma's wedding can only take place after the theft of Mrs. Weston's turkeys—not so far from the poultry-yard after all, if only by association.

31. For an exhaustive reading of this conversation, see Gabrielle D.V. White, *Jane Austen in the Context of Abolition: "A Fling at the Slave Trade"* (New York: Palgrave Macmillan, 2006), 60–65.

32. An Act for the Abolition of the Slave Trade was passed on March 25, 1807. It abolished slave trade in the British empire although not the act of slavery itself. Galperin suggests that the allusion to the slave trade Mrs. Elton makes in this scene may account for the source of her family's recent wealth.

33. Thompson, *Models of Value*, 185–186.

34. For a fuller account of the Austen family's efforts to produce this image of Jane Austen, see Deirdre Le Faye, *A Chronology of Jane Austen* (Cambridge and New York: Cambridge University Press, 2006).

35. Miles Ogborn, *Indian Ink: Script and Print in the Making of the English East India Company* (Chicago and London: University of Chicago Press, 2007), 21.

36. C.A. Bayly, *Empire and Information: Intelligence Gathering and Social Communication in India, 1780–1870* (Cambridge: Cambridge University Press, 1996), 76–77.

37. For an illuminating account of the self-consciously produced archive of British colonialism, and the complicated place women had in shaping the narrative of empire, see Betty Joseph, *Reading the East India Company 1720–1840: Colonial Currencies of Gender* (Chicago: University of Chicago Press, 2004).

EPILOGUE

1. *Philosophical Transactions*, 599.
2. Ibid., 604–605.

3. Although Werrett also claims that the "localist" model has been crucial to nuancing accounts of colonial science and technology. See David Chambers and Richard Gillespie, "Locality in the History of Science: Colonial Science, Technoscience, and Indigenous Knowledge," *Osiris* 15 (2000): 21–40.

4. Simon Werrett, "Technology on the Spot: The Trials of the Congreve Rocket in India in the Early Nineteenth Century," *Technology and Culture* 53 (July 2012): 600.

5. Ibid.

6. Ibid.

7. Or so claims Dava Sobel, the author of the popular history *Longitude*. Richard Dunn and Rebekah Higgit, however, argue that Shovell's accident didn't have as much to do with the passing of the Longitude Act as Sobel represents. Whatever the case, the fact remains that Shovell's miscalculations demonstrated an enormous problem in nautical navigation: how does one effectively determine longitude at sea? See Dava Sobel, *Longitude: The True Story of a Lone Genius Who Solved the Greatest Scientific Problem of His Time* (London: Walker & Company, 1995) and Richard Dunn and Rebekah Higgit, *Finding Longitude: How Ships, Clocks, and Stars Help Solve the Longitude Problem* (London: Collins, 2014).

8. Cathy Caruth, "Unclaimed Experience: Trauma and the Possibility of History," *Yale French Studies* 79 (1991): 182.

9. For another argument on longitude and the establishment of the British empire, see Bernhard Siegert, "Longitude and Simultaneity in Philosophy, Physics, and Empires," *Configurations* 23:2, 145–163.

10. Nevile Maskelyne was a major figure in retaining lunar calculations over Harrison's marine chronometers. See Sobel, *Longitude*, 197–199.

11. The *Oxford English Dictionary* defines trilateration as "A method of surveying analogous to triangulation in which each triangle is determined by the measurement of all three sides." Although this definition is largely geometric, trilateration has practical applications in surveying and navigation, including the Global Positioning System.

12. Defoe, *An Account of Monsieur De Quesne's Late Expedition at Chio Together with the Negotiation of Monsieur Guilleragues, the French Ambassador at the Port/ in a Letter Written by an Officer of the Grand Vizir's to a Pacha; translated into English* (1683), Eighteenth Century Collections Online.

13. Alexander Dalrymple, *An account of the discoveries made in the South Pacifick Ocean, previous to 1764. Part I. Containing, I. A Geographical Description of Places. II. An Examination of the Conduct of the Discoverers in the Tracks they pursued. III. Investigations of what may be further expected.* London, Printed in the Year 1767 [1769], Eighteenth Century Collections Online.

14. Sobel, *Longitude*, 162.

15. Ibid., 197–199.

16. I refer here to Benedict Anderson's invaluable formulation in which he argues "the nation is always conceived as a deep, horizontal comradeship. Ultimately it is this fraternity that makes it possible, over the past two centuries, for so many millions of people, not so much to kill, as willingly to die for such limited imaginings." See Benedict Anderson, *Imagined Communities: Reflections on the Origin and Spread of Nationalism* (London: Verso, 1983), 7.

17. C.F. Volney, *The Ruins, or Meditation on the Revolutions of Empire: And, the Law of Nature*, Project Gutenberg, http://www.gutenberg.org/ebooks/1397, ch. II.

18. See John Locke, *An Essay Concerning Human Understanding. In Four Books* (London, printed by Eliz. Holt for Thomas Basset, 1690).

19. Poovey rests her claim on double-entry bookkeeping, which first emerges in England in 1588, and argues that the notion of balance that is also a product of the fictions of "money" and "price" "produced the system's most salient meaning: that the merchant who kept the books obeyed the order of God's harmonious world, that the merchant was creditworthy because he was honest." See Mary Poovey, *A History of the Modern Fact: Problems of Knowledge in the Sciences of Wealth and Society* (Chicago and London: University of Chicago Press, 1998), 11.

Agamben, Giorgio. *Homo Sacer: Sovereign Power and Bare Life.* Trans. Daniel Heller-Roazen. Stanford: Stanford University Press, 1998.

Anderson, Benedict. *Imagined Communities: Reflections on the Origin and Spread of Nationalism.* London: Verso, 1983.

Andrea, Bernadette. *Women and Islam in Early Modern English Literature.* Cambridge: Cambridge University Press, 2007.

———. "Alternatives to Orientalism?: Mary Wortley Montagu and Her 'Turkish' Son." In *Britain and the Muslim World: Historical Perspectives,* ed. Gerald Maclean. Newcastle: Cambridge Scholars, 2011.

Aravamudan, Srinivas. *Tropicopolitans: Colonialism and Agency, 1688–1804.* Durham and London: Duke University Press, 1999.

———. *Enlightenment Orientalism: Resisting the Rise of the Novel.* Chicago and London: University of Chicago Press, 2012.

Arbuthnot, John. *An Essay Concerning the Effects of Air on Human Bodies.* 1733. Eighteenth Century Collections Online.

Armstrong, Nancy. *Desire and Domestic Fiction: A Political History of the Novel.* New York: Oxford University Press, 1987.

Arnold, David. *Colonizing the Body: State Medicine and Epidemic Disease in Nineteenth-Century India.* Berkeley and London: University of California Press, 1993.

Austen, Jane. *Emma.* Eds. Richard Cronin and Dorothy McMillan. Cambridge: Cambridge University Press, 2005.

Baber, Zaheer. *The Science of Empire: Scientific Knowledge, Civilization, and Colonial Rule in India.* Albany: State University of New York Press, 1996.

Badiou, Alain. *Being and Event.* Trans. Oliver Fetham. London and New York: Continuum, 2005.

Banthia, Jayant, and Tim Dyson, "Smallpox in Nineteenth-Century India" *Population and Development Review* 25:4 (December 1999): 649–680.

Barchas, Janine. *Matters of Fact in Jane Austen: History, Location, and Celebrity.* Baltimore: Johns Hopkins University Press, 2012.

Barrell, John. *The Infection of Thomas De Quincey.* New Haven: Yale University Press, 1991.

Barros, Carolyn A., and Johanna M. Smith, *Life Writings by British Women, 1660–1850: An Anthology.* Boston: Northeastern University Press, 2000.

Bayly, C.A. *Empire and Information: Intelligence Gathering and Social Communication in India, 1780–1870.* Cambridge: Cambridge University Press, 1996.

Behn, Aphra. *Oroonoko.* Ed. Janet Todd. New York and London: Penguin, 2003.

Berne, Eric. "The Psychological Structures of Space with Some Remarks on *Robinson Crusoe.*" *Psychological Quarterly* 25 (1956): 549–567.

Bewell, Alan. *Romanticism and Colonial Disease* Cambridge: Cambridge University Press, 1999.

Bingham, Hiram. *Elihu Yale: Governor, Collector, Benefactor.* Proceedings: American Antiquarian Society, Vol. 47: 93–144. Worcester, 1937.

———. *Elihu Yale: The American Nabob of Queen Square.* New York: Dodd, Mead & Company, 1939.

Boxer, Charles Ralph. *From Lisbon to Goa, 1500–1750: Studies in Portuguese Maritime Enterprise.* London: Variorum Reprints, 1984.

Brady, Frank, and W.K. Wimsatt, eds. *Samuel Johnson: Selected Poetry and Prose.* Berkeley: University of California Press, 1977.

Braudel, Fernand. *The Structures of Everyday Life.* 2 vols. New York: Harper and Row, 1979.

Brontë, Charlotte. *Jane Eyre.* Ed. Richard J. Dunne. New York and London: Norton, 1993.

Brown, Laura. *Alexander Pope.* New York and Oxford: Basil Blackwell, 1985.

———. *Ends of Empire: Women and Ideology in Early Eighteenth-Century English Literature.* Ithaca: Cornell University Press, 1993.

———. *Fables of Modernity: Literature and Culture in the English Eighteenth Century.* Ithaca, NY: Cornell University Press, 2001.

Brownson, Orestes A. *The Everlasting Yes.* In *The Transcendentalists: An Anthology*, ed. Perry Miller. Cambridge, MA and London: Harvard University Press, 1977.

Burrell, David B. *Analogy and Philosophical Language.* New Haven: Yale University Press, 1973.

Carpue, Joseph Constantine. *An Account of Two Successful Operations for Restoring a Lost Nose from the Integuments of the Forehead . . . Including Descriptions of the Indian and Italian Methods.* London: Longman et al., 1816.

Caruth, Cathy. "Unclaimed Experience: Trauma and the Possibility of History." *Yale French Studies* 79 (1991): 181–192.

Chakrabarty, Dipesh. *Provincializing Europe: Postcolonial Thought and Historical Difference*. Princeton and Oxford: Princeton University Press, 2000.

Chaplin, Joyce E. *Subject Matter: Technology, the Body, and Science on the Anglo-American Frontier, 1500–1676*. Cambridge: Harvard University Press, 2001.

Chaudhuri, K.N. *Asia before Europe: Economy and Civilisation of the Indian Ocean from the Rise of Islam to 1750*. Cambridge: Cambridge University Press, 1990.

Cleland, John. *Fanny Hill, or Memoirs of a Woman of Pleasure*. Ed. Peter Wagner. New York: Penguin, 1985.

Colley, Linda. *Captives: The Story of Britain's Pursuit of Empire and How Its Soldiers and Civilians Were Held Captive by the Dream of Global Supremacy, 1600–1850*. New York: Pantheon, 2002.

Cooper, Frederick. "What Is the Concept of Globalization Good for?: An African Historian's Perspective." *African Affairs* 100: 399 (April 2001): 189–213.

Copeland, Edward. *Women Writing about Money: Women's Fiction in England, 1790–1820*. Cambridge and New York: Cambridge University Press, 1995.

———. "Money." In *The Cambridge Companion to Jane Austen*, ed. Edward Copeland and Juliet McMaster. Cambridge: Cambridge University Press, 2010; online 2011. http://dx.doi.org/10.1017/CCO9780521763080.

Dalrymple, Alexander. *An account of the discoveries made in the South Pacifick Ocean, previous to 1764. Part I. Containing, I. A Geographical Description of Places. II. An Examination of the Conduct of the Discoverers in the Tracks they pursued. III. Investigations of what may be further expected*. London, Printed in the Year 1767 [1769]. Eighteenth Century Collections Online.

Dames, Nicholas. *Amnesiac Selves: Nostalgia, Forgetting, and British Fiction, 1810–1870*. Oxford: Oxford University Press, 2001.

de Certeau, Michel. *The Practice of Everyday Life*. Trans. Steven Randall. Berkeley: University of California Press, 1984.

De Quincey, Thomas. *Confessions of an English Opium Eater*. Ed. Grevel Lindop. Oxford: Oxford University Press, 1985.

Debus, Allen. *The Chemical Philosophy: Paracelsian Science and Medicine in the Sixteenth and Seventeenth Centuries*. Vol. I. New York: Science History Publications, 1977.

Defoe, Daniel. *Robinson Crusoe*. Ed. Thomas Keymer. Oxford: Oxford University Press, 2007.

———. *Journal of the Plague Year*. Ed. Louis Landa. Oxford: Oxford University Press, 2010.

———. *An Account of Monsieur De Quesne's Late Expedition at Chio Together with the Negotiation of Monsieur Guilleragues, the French Ambassador at the Port/ in a Letter Written by an Officer of the Grand Vizir's to a Pacha; translated into English* (1683). Eighteenth Century Collections Online.

Dharampal. *The Beautiful Tree: Indigenous Indian Education in the Eighteenth Century*. Vol. III. Goa: Other India Press, 2000.

DiPiero, Thomas. "Unreadable Novels: Toward a Theory of Seventeenth-Century Aristocratic Fiction." *Novel: A Forum on Fiction* 38:2–3 (Spring /Summer 2005): 129–146.

Downie, J.A. "Who Says She's a Bourgeois Writer?: Reconsidering the Social and Political Contexts of Jane Austen's Novels." *Eighteenth-Century Studies* 40:1 (2006): 69–84.

Dunn, Richard, and Rebekah Higgit. *Finding Longitude: How Ships, Clocks, and Stars Help Solve the Longitude Problem*. London: Collins, 2014.

Durbach, Nadja. *Bodily Matters: The Anti-Vaccination Movement in England, 1853–1907*. Durham and London: Duke University Press, 2005.

Epstein, William H. *John Cleland: Images of a Life*. New York and London: Columbia University Press, 1974.

Ermarth, Elizabeth. *Realism and Consensus in the English Novel: Time, Space, and Narrative*. Princeton: Princeton University Press, 1998.

Ezell, Margaret J.M. *Writing Women's Literary History*. Baltimore: Johns Hopkins University Press, 1993.

Fagan, Brian. *The Little Ice Age: How Climate Made History, 1300–1850*. New York: Basic Books, 2000.

Fergus, Jan. *Jane Austen: A Literary Life*. Great Britain: Macmillan, 1991.

Ferguson, Niall. *Empire: The Rise and Demise of the British World Order and the Lessons for Global Power*. New York: Basic Books, 2002.

Forbes, John. *Cyclopaedia of Practical Medicine*. Vol. 3. London: Whittaker, Treacher & Co., 1834.

Frank, Andre Gunder. *Re-Orient: Global Economy in the Asian Age*. Berkeley and London: University of California Press, 1998.

Galperin, William. H. *The Historical Austen*. Philadelphia: University of Pennsylvania Press, 2003

Gilbert, Sandra and Susan Gubar. *The Madwoman in the Attic*. New Haven and London: Yale University Press, 1979.

Gilman, Sander. *Disease and Representation: Stereotypes of Sexuality, Race, and Madness*. Ithaca: Cornell University Press, 1985.

Gladfelder, Hal. *Fanny Hill in Bombay: The Making and Unmaking of John Cleland*. Baltimore: Johns Hopkins University Press, 2012.

Goody, Jack. *The East in the West*. Cambridge: Cambridge University Press, 1996.

Greenblatt, Stephen. *Marvelous Possessions: The Wonder of the New World*. Chicago: University of Chicago Press, 1991.

Grossman, Joyce. "Ayahs, Dhayes, and Bearers: Mary Sherwood's Indian Experience and Constructions of Subordinated Others," *South Atlantic Review* 66:2 (Spring 2001): 14–44.

Grove, Richard H. *Green Imperialism: Colonial Expansion, Tropical Island Eden's and the Origins of Environmentalism, 1600–1660*. Cambridge: Cambridge University Press, 1995.

Gupta, Anil K., David M. Anderson, and Jonathan T. Overpeck. "Abrupt Changes in the Asian Southwest Monsoon during the Holocene and Their Links to the North Atlantic Ocean." *Nature* 421 (January 23, 2003): 354–357.

Habermas, Jürgen. *The Structural Transformation of the Public Sphere: An Inquiry into a Category of Bourgeois Society*. Trans. Thomas Burger and Frederick Lawrence. Cambridge: MIT Press, 1989.

Hill, Christopher. *Century of Revolution 1603–1714*. Vol. 5. Edinburgh: T. Nelson, 1961.

Herbert, Eugenia W. "Smallpox Inoculation in Africa." *Journal of African History* 16:4 (1975): 539–559.

Heyne, Benjamin. *Tracts, historical and statistical, on India: with journals of several tours through various parts of the peninsula: also, an account of Sumatra, in a series of letters*. London: Robert Baldwin, and Black, Perry, and Co., 1814.

Hunter, W.W. *History of British India*. Vol. II. New York and Bombay: Longmans, Green and Co., 1912.

Jenkins, Eugenia Zuroski. *A Taste for China: English Subjectivity and the Prehistory of Orientalism*. New York and Oxford: Oxford University Press, 2013.

Johnson, Claudia L. *Jane Austen: Women, Politics, and the Novel*. Chicago and London: University of Chicago Press, 1988.

Johnson, Samuel. "Preface to the Dictionary. 1755." In *Samuel Jackson: Selected Poetry and Prose*, ed. Frank Brady and W.K. Wimsatt. Berkeley: University of California Press, 1977.

———. "Rasselas." In *Samuel Jackson: Selected Poetry and Prose*, ed. Frank Brady and W.K. Wimsatt. Berkeley: University of California Press, 1977.

———. "Vanity of Human Wishes." In *Samuel Jackson: Selected Poetry and Prose*, ed. Frank Brady and W.K. Wimsatt. Berkeley: University of California Press, 1977.

Joseph, Betty. *Reading the East India Company, 1720–1840: Colonial Currencies of Gender.* Chicago: University of Chicago Press, 2004.

Kaul, Suvir. *Poems of Nation, Anthems of Empire: English Verse in the Long Eighteenth Century.* Charlottesville and London: University Press of Virginia, 2000.

Keay, John. *The Honourable Company: A History of the English East India Company.* New York: Macmillan, 1991.

Kindersley, Jemima. *Letters from the Island of Teneriffe, Brazil, the Cape of Good Hope, and the East Indies.* In *Life Writings by British Women, 1660–1850: An Anthology,* ed. Carolyn Barros and Johanna Smith. Boston: Northeastern University Press, 2000.

Kipling, Rudyard. *Poems.* Ed. Peter Washington. New York: Everyman, 2013.

Klein, Ursula, and Wolfgang Lefèvre. *Materials in Eighteenth-Century Science: A Historical Ontology.* Cambridge and London: MIT Press, 2007.

Knox-Shaw, Peter. *Jane Austen and the Enlightenment.* Cambridge: Cambridge University Press, 2004.

Kolko, Beth E., Lisa Nakamura, and Gilbert B. Rodman, eds. *Race in Cyberspace.* New York and London: Routledge, 2000.

Kristeva, Julia. *Powers of Horror: An Essay on Abjection.* Trans. Leon S. Rodriguez. New York: Columbia University Press, 1982.

Kuhn, Thomas. *The Structure of Scientific Revolutions.* Chicago: University of Chicago Press, 1962.

Lacan, Jacques. *Écrits.* Trans. Alan Sheridan. New York and London: Norton, 1981.

———. *The Four Fundamental Concepts of Psycho-Analysis.* Trans. Alan Sheridan. New York and London: Norton, 1981.

Ladurie, Emmanuel Le Roy. *Times of Feast, Times of Famine: A History of Climate since the Year 1300.* Trans. Barbara Bray. London: George Allen & Unwin, 1971.

Lamb, Jonathan. *Preserving the Self in the South Seas, 1680–1840.* Chicago: University of Chicago Press, 2001.

———. *The Things Things Say.* Princeton and Oxford: Princeton University Press, 2011.

Landry, Donna. *The Invention of the Countryside: Hunting, Walking, and Ecology in English Literature, 1671–1831.* New York, Palgrave, 2001.

Latour, Bruno. *We Have Never Been Modern.* Trans. Catherine Porter. Cambridge: Harvard University Press, 1993.

———. *Pandora's Hope: Essays on the Reality of Science Studies.* Cambridge and London: Harvard University Press, 1999.

Le Faye, Deirdre. *A Chronology of Jane Austen.* Cambridge and New York: Cambridge University Press, 2006.

Leask, Nigel. *British Romantic Writers and the East: Anxieties of Empire.* Cambridge: Cambridge University Press, 1992.

Liu, Lydia H. "Robinson Crusoe's Earthenware Pot." *Critical Inquiry* 25:4 (Summer 1999): 728–757.

Locke, John. *An Essay Concerning Human Understanding. In Four Books.* London, Printed by Eliz. Holt for Thomas Basset, 1690.

Love, Henry Davison. *Vestiges of Old Madras, 1640–1800, Traced from the East India Company's Records Preserved at Fort St. George and the India Office and from Other Sources.* Vol. I: *1913.* New York: AMS Press, 1968.

Makdisi, Saree. *Romantic Imperialism: Universal Empire and the Culture of Modernity.* Cambridge: Cambridge University Press, 1998.

Markley, Robert. *Fallen Languages: Crises of Representation in Newtonian England, 1660–1740.* Ithaca: Cornell University Press, 1993.

———. *The Far East and the English Imagination, 1600–1730.* Cambridge: Cambridge University Press, 2006.

———. "'A Putridness in the Air': Monsoons and Mortality in Seventeenth-Century Bombay." *Journal for Early Modern Cultural Studies* 10:2 (Fall/Winter 2010): 105–125.

McDowell, Paula. *The Women of Grub Street: Press, Politics, and Gender in the Literary Marketplace, 1678–1730.* Oxford and New York: Oxford University Press, 1998.

Milton, Giles. *Nathaniel's Nutmeg or, The True and Incredible Adventures of the Spice Trader Who Changed the Course of History.* New York: Penguin, 1999.

Miskin, Lauren. "'True Indian Muslin' and the Politics of Consumption in Jane Austen's *Northanger Abbey.*" *Journal for Early Modern Cultural Studies* 15:2 (Spring 2015): 5–26.

Mitchell, W.T.J. "Picturing Terror: Derrida's Autoimmunity." *Critical Inquiry* 33:2 (2007): 277–290.

Monaghan, David, ed. *Jane Austen in a Social Context.* Totowa, NJ: Barnes and Noble Books, 1981.

Montagu, Lady Mary Wortley, *Letters from Constantinople.* In *Embassy to Constantinople: The Travels of Lady Mary Wortley Montagu,* ed. Christopher Pick. New York: New Amsterdam Books, 1988.

Nash, Richard. *Wild Enlightenment: The Borders of Human Identity in the Eighteenth Century.* Charlottesville and London: University of Virginia Press, 2003.

Neill, Edward. *The Politics of Jane Austen.* London and New York: Macmillan, 1999.

Nightingale, Carl. "Madras, New York, and the Urban and Global Origins of Color Lines." Unpublished manuscript in author's possession (courtesy Carl Nightingale).

Nussbaum, Felicity A. *Torrid Zones: Maternity, Sexuality, and Empire in Eighteenth-Century English Narratives*. Baltimore: Johns Hopkins University Press, 1995.

Ogborn, Miles. *Indian Ink: Script and Print in the Making of the English East India Company*. Chicago and London: University of Chicago Press, 2007.

Orwell, George. *Burmese Days*. Orlando: Harcourt Brace Jovanovich, 1962.

Philip, Kavita. *Civilizing Natures: Race, Resources, and Modernity in Colonial South India*. New Brunswick, NJ: Rutgers University Press, 2004.

Pomeranz, Kenneth. *The Great Divergence: Europe, China, and the Making of the Modern World Economy*. Princeton: Princeton University Press, 2000.

Pope, Alexander. *Windsor-Forest*. In *Poetry and Prose of Alexander Pope*. Ed. Aubrey Williams. Boston: Houghton Mifflin, 1969.

———. *The Rape of the Lock*. Ed. Cynthia Wall. Boston and New York: Bedford, 1998.

Pollak, Ellen. *The Poetics of Sexual Myth: Gender, and Ideology in the Verse of Swift and Pope*. Chicago and London: University of Chicago Press, 1985.

Pollak, Kurt, and E. Ashworth Underwood. *The Healers: The Doctor, Then and Now*. London: Thomas Nelson and Sons, 1963.

Pollock, Anthony. *Gender and the Fictions of the Public Sphere, 1690–1755*. New York: Routledge, 2009.

Poovey, Mary. *A History of the Modern Fact: Problems of Knowledge in the Sciences of Wealth and Society*. Chicago and London: University of Chicago Press, 1998.

Porter, David. *The Chinese Taste in Eighteenth-Century England*. Cambridge: Cambridge University Press, 2010.

Prakesh, Om. "British Private Traders in the Indian Ocean in the Seventeenth and Eighteenth Centuries." In *Britain and the Muslim World: Historical Perspectives*, ed. Gerald Maclean. Newcastle upon Tyne: Cambridge Scholars Publishing, 2011.

Reid, Anthony. *Southeast Asia in the Age of Commerce, 1450–1680*. New Haven: Yale University Press, 1988.

Reiss, Timothy J. *The Discourse of Modernism*. Ithaca and London: Cornell University Press, 1982.

Richetti, John. "Robinson Crusoe: The Self as Master." In *Robinson Crusoe: An Authoritative Text, Context, Criticism*. New York: Norton, 1994.

Rotman, Brian. *Signifying Nothing: The Semiotics of Zero*. Stanford: Stanford University Press, 1987.

Rusnock, Andrea. *Vital Accounts: Quantifying Health and Population in Eighteenth-Century England and France*. Cambridge: Cambridge University Press, 2002.

———. "Catching Cowpox: The Early Spread of Smallpox Vaccinations, 1798–1810." *Bulletin of the History of Medicine* 83 (2009): 17–36.

Rzepka, Charles. *Sacramental Commodities: Gift, text, and the Sublime in De Quincey*. Amherst: University of Massachusetts Press, 1995.

Schellenberg, Betty A. *The Professionalization of Women Writers in Eighteenth-Century Britain*. Cambridge and New York: Cambridge University Press, 2005.

Schleifer, Ronald. *Analogical Thinking: Post-Enlightenment Understanding in Language, Collaboration, and Interpretation*. Ann Arbor: University of Michigan Press, 2000.

Scott, James C. *Seeing Like a State: How Certain Schemes to Improve the Human Condition Have Failed*. New Haven: Yale University Press, 1998.

———*The Art of Not Being Governed: An Anarchist History of Upland Southeast Asia*. New Haven and London: Yale University Press, 2009.

Shapin, Steven. *A Social History of Truth: Civility and Science in Seventeenth-Century England*. Chicago: University of Chicago Press, 1994.

Shapin, Steven, and Simon Schaffer. *Leviathan and the Air Pump: Hobbes, Boyle, and the Experimental Life*. Princeton: Princeton University Press, 1985.

Sharpe, Jenny. *Allegories of Empire: The Figure of the Woman in the Colonial Text*. Minneapolis: University of Minnesota Press, 1993.

Shuttleworth, Sally. "Female Circulation: Medical Discourse and Popular Advertising in the Mid-Victorian Era." In *Body Politics: Women and the Discourses of Science*, ed. Mary Jacobus, Evelyn Fox Keller, Sally Shuttleworth. New York and London: Routledge.

Siegert, Bernhard. "Longitude and Simultaneity in Philosophy, Physics, and Empires." *Configurations* 23:2 (2012): 144–163.

Smith, Chloe Wigston. "'Callico Madams': Servants, Consumption, and the Calico Crisis." *Eighteenth-Century Life* 31:2 (2007): 29–55.

Smith, Goldwin. *The Life of Jane Austen*. London: W. Scott, 1890.

Smith, Pamela H. *The Business of Alchemy: Science and Culture in the Holy Roman Empire*. Princeton: Princeton University Press, 1994.

Sobel, Dava. *Longitude: The True Story of a Lone Genius Who Solved the Greatest Scientific Problem of His Time*. London: Walker & Company, 1995.

Spary, E.C. *Utopia's Garden: French Natural History from Old Regime to Revolution*. Chicago: University of Chicago Press, 2000.

———. "Of Nutmegs and Botanists: The Colonial Cultivation of Botanical Identity." In *Colonial Botany: Science, Commerce, and Politics in the Early*

Modern World, ed. Londa Schiebinger and Claudia Swan. Philadelphia: University of Pennsylvania Press, 2005

Stewart, Maaja A. *Domestic Realities and Imperial Fictions: Jane Austen's Novels in Eighteenth-Century Contexts*. Athens and London: University of Georgia Press, 1993.

Stoker, Bram. *Dracula*. Eds. Nina Auerbach and David Skal. New York: Norton, 1997.

Sudan, Rajani. "Lost in Lexicography: Legitimating Cultural Identity in Johnson's *Preface* to the *Dictionary*." *ECTI* 39:2 (1998): 127–146.

———. *Fair Exotics: Xenophobic Subjects in English Literature, 1720–1850*. Philadelphia: University of Pennsylvania Press, 2002.

Suleri, Sara. *The Rhetoric of English India*. Chicago and London: University of Chicago Press, 1992.

Tebb, William. "Sanitation, Not Vaccination: The True Protection against Small-Pox." Paper presented at the Second International Congress of Anti-Vaccinators, Cologne, October 12, 1881.

Thompson, James. *Models of Value: Eighteenth-Century Political Economy and the Novel*. Durham and London: Duke University Press, 1996.

Thoreau, Henry David. *Natural History of Massachusetts*. In *The Transcendentalists: An Anthology*, ed. Perry Miller. Cambridge, MA and London: Harvard University Press, 1977.

———. *Journal*. Vol. 2: *1842–1848*. Ed. Robert Sattelmeyer. Princeton: Princeton University Press, 1984.

———. *Walden*. Ed. Jeffrey S. Cramer. New Haven and London: Yale University Press, 2004.

Travers, Robert. *Ideology and Empire: The British in Bengal*. New York and Cambridge: Cambridge University Press, 2007.

Van Renen, Denys. "Montagu's Letters from the Levant: Contesting the Borders of European Selfhood." *Journal for Early Modern Cultural Studies* 11:2 (Fall 2011): 1–34.

Viswanathan, Gauri. "The Naming of Yale College." In *Cultures of United States Imperialism*, ed. Amy Kaplan and Donald E. Pease. Durham and London: Duke University Press, 1993.

Volney, C.F. *The Ruins, or Meditation on the Revolutions of Empire: And the Law of Nature*. Project Gutenberg. http://www.gutenberg.org/ebooks /1397.

Wang, R. Bin. *The Asian Monsoon*. Chichester, UK: Praxis, 2006.

Werrett, Simon. "Technology on the Spot: The Trials of the Congreve Rocket in India in the Early Nineteenth Century." *Technology and Culture* 53 (July 2012): 508–624.

Wheeler, J. Talboys. *Early Records of British India: A History of the English Settlements in India*. London: Curzon Press, 1878.

White, Gabrielle D.V. *Jane Austen in the Context of Abolition: "A Fling at the Slave Trade."* New York: Palgrave Macmillan, 2006.

White, Richard. "The Nationalization of Nature." *Journal of American History*, 86:3 (1999): 976–986.

Woolf, Virginia.*The Common Reader*. New York: Harcourt, 1925.

———. *A Room of One's Own*. New York: Harcourt Brace Jovanovich, 1929.

Wrightson, Keith. *Earthly Necessities: Economic Lives in Early Modern Britain*. New Haven and London: Yale University Press, 2000.

Yang, Chi-Ming. *Performing China: Virtue, Commerce, and Orientalism in Eighteenth-Century England, 1660–1760*. Baltimore: Johns Hopkins University Press, 2011.